The Gut in Parkinson's Disease

George Triadafilopoulos

The Gut in Parkinson's Disease

Springer

George Triadafilopoulos
Division of Gastroenterology and Hepatology
Stanford University School of Medicine
Stanford, CA, USA

ISBN 978-3-031-77170-5 ISBN 978-3-031-77171-2 (eBook)
https://doi.org/10.1007/978-3-031-77171-2

This Springer imprint is published by the registered company Springer Nature Switzerland AG
The registered company address is: Gewerbestrasse 11, 6330 Cham, Switzerland

For Runa Lascar, MD, FACP

Le chaleur de ma vie

About this Book

Gastroenterology is a medical subspecialty, that beyond endoscopy and the management of gastrointestinal conditions, has taken an additional focus: the management of gut health in the presence of other diseases that predominantly affect other organs, such as Parkinson's Disease (PD), and understanding how they impact the gut.

Within this construct, Neurogastroenterology addresses any gastrointestinal dysmotility and altered sensation in patients with Parkinson's Disease that may occur at different phases in its evolution, during the prodromal state (i.e., increasingly disabling constipation), during the active disease phase, which may last for decades, and particularly during its various treatment phases (i.e., pharmacologic therapy, endoscopic interventions, or surgery). As such, it focuses on the diagnosis and management of such patients to optimize their clinical care while maximizing their quality of life in an interdisciplinary cooperation with other specialists, particularly those managing movement disorders.

This book is designed to help general gastroenterologists, primary care physicians, neurologists, movement disorders specialists, and their allied teams take care of their patients with gastrointestinal problems in the context of Parkinson's Disease that are not straightforward to master. The text is written from the standpoint of an expert gastroenterologist who practiced and conducted research in gastrointestinal dysmotility and altered sensation in patients with Parkinson's Disease for over 20 years.

The main objective of this book is to compile and present the most clinically important issues in Neurogastroenterology of the gut in Parkinson's Disease (Gut-PD) to help anyone taking care of these patients' gastrointestinal problems. While the text may be targeted particularly toward practicing general gastroenterologists, it certainly can be of benefit to gastroenterologists and neurologists in training, general internists and movement disorders specialists as well as advanced practice providers. It can also be a useful guide to patients with Parkinson's Disease and their families to facilitate exchanges with their clinical team and improve their understanding of the disease.

Contents

Introduction

1

Parkinson's Disease (PD): Generalities

Parkinson's disease (PD) is a progressive degenerative disease of the nervous system that affects over one million people in North America. This number is expected to rise to 1.2 million by 2030. Patients suffering from PD mainly present with motor (movement) difficulties, such as tremors, slow movement of the body and extremities, and muscle rigidity, but other nerve, psychiatric, and non-motor manifestations may occur. The incidence and prevalence of PD increase with age, with a male-to-female ratio of about 2:1 [1]. The economic burden of Parkinson's disease in the USA is predicted to increase $79 billion in 2037 [2] (Fig. 1.1).

PD is part of a group of neurodegenerative disorders collectively known as synucleinopathies, which are characterized by the deposition of phosphorylated α-synuclein (P-SYN) within the central and peripheral nervous systems. These

Key points in Parkinson's Disease
A progressive multisystem neurological disorder of later life
Motor features related to degenerationof dopaminergic brain-stem neurons
Commonly associated with disordered mood, sleep, sensation, cognition, and autonomic function
20% genetic
No therapy slows progression
Comprehensive multidisciplinary care is essential

Fig. 1.1 Key points in Parkinson's Disease (PD) discussed in this introductory chapter

G. Triadafilopoulos, *The Gut in Parkinson's Disease*,
https://doi.org/10.1007/978-3-031-77171-2_1

include the Lewy body diseases (Parkinson's disease [PD], dementia with Lewy bodies [DLB]), multiple system atrophy (MSA), and pure autonomic failure (PAF) that have overlapping clinical features, progressive neurodegeneration, and disability [3, 4].

PD and dementia with Lewy bodies are presently clinically defined, with detection of α-synuclein pathology used as the gold standard for the diagnosis. Over the years, many clinical criteria, such as the Hoehn and Yahr scale, have been proposed to describe disease severity. However, the clinical syndromes and disease progression are heterogeneous and overlapping, while biomarkers, clinical, epidemiological, and neuropathological data have increasingly shown a prolonged period of neurodegeneration that begins long before any symptoms or signs. The use of biomarkers that measure misfolded, predominantly neuronal α-synuclein (n-αsyn), the neuropathological hallmark of PD and dementia with Lewy bodies, could be valuable. Until recently, n-αsyn could only be reliably measured at autopsy, but the rapid validation of a seed amplification assay, which could detect n-αsyn in vivo with high accuracy, has been a landmark advancement calling for a redefinition of PD and dementia with Lewy bodies based on biology rather than clinical presentation [5] (Fig. 1.2).

Recently, a biological definition has been proposed that combines PD and dementia with Lewy bodies under the term neuronal α-synuclein disease, based on in vivo detection of n-αsyn. Such individuals with n-αsyn are at high risk for developing dopaminergic neuronal dysfunction, in turn enabling biology-specific therapeutic development and intervention before symptom onset to potentially prevent or halt progression and to better identify and biologically define the disease. The neuronal α-synuclein disease integrated staging system (NSD-ISS) has been recently

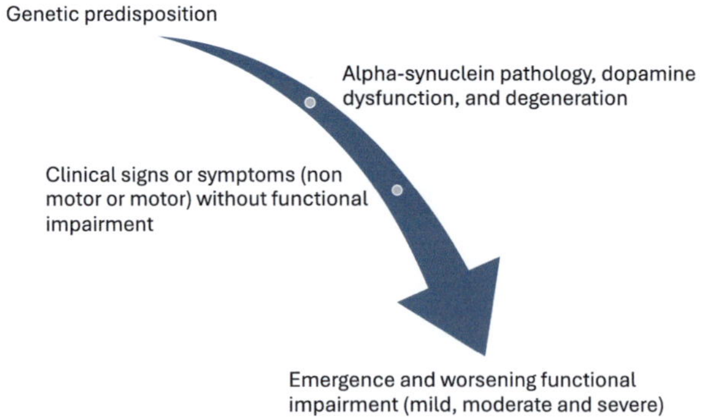

Fig. 1.2 Outline of the evolution of PD, starting with genetic predisposition, to neuronal degeneration, to symptoms and, ultimately, functional impairment

introduced as a biological and clinical staging system for neuronal α-synuclein disease that starts while the process is still asymptomatic and continues through the development of increasing dysfunction and clinical impairment and is aimed at including in ongoing research studies participants without symptoms but with specific biomarker profiles, as well as those in later stages, with evolving disease manifestations [6].

An estimated 2.5 million people in the USA have a diagnosis of a synucleinopathy, and approximately 180,000 new patients are diagnosed each year [7]. No disease-modifying therapy exists for these diseases [8]. Many patients with synucleinopathy experience diagnostic delay or an initial misdiagnosis even when evaluated at movement disorder centers [9, 10].

Parkinson's Disease (PD): Two Groups of Diseases

The following diagram distinguishes PD based on two key elements: (1) the well-known motor disease, with its classic as well as atypical neurological manifestations, such as muscle rigidity, bradykinesia, and tremor, and (2) the non-motor disease, with hyposmia, gut manifestations (gut PD), hypotension, altered cognition, and sleep. This wide spectrum of clinical presentations and their differential and evolving effects on quality of life (QoL) require expert simultaneous assessment, not only by neurologists specializing in movement disorders, but also autonomic function experts, neurogastroenterologists, endoscopists, and surgeons (Fig. 1.3).

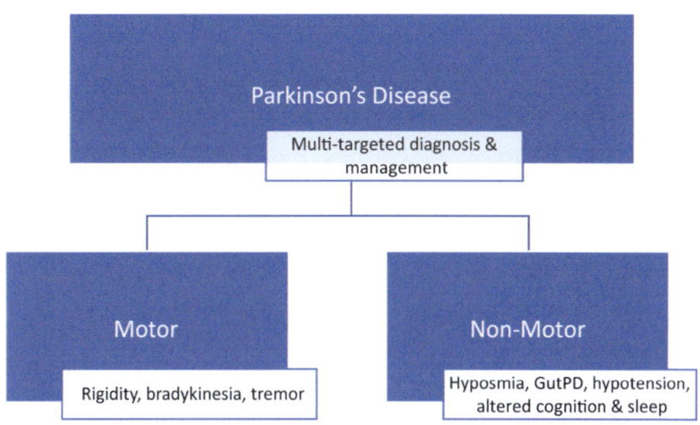

Fig. 1.3 Diagram depicting the motor and non-motor elements in PD

Parkinson's Disease and the Gut

Gastrointestinal (GI) dysfunction is one of the most common non-motor symptoms in PD, occurring in about 60–80% of patients. These include constipation, evacuation disorders, dysphagia, gastroparesis, and small intestinal bacterial overgrowth. Indeed, the GI tract is thought to have an important role in the pathophysiology of PD [11].

If one dissects the brain of patients with PD, they will find the Lewy bodies or Lewy neurites, characterized by abnormal accumulation of a protein, called α-synuclein. Similarly, abnormal accumulation of α-synuclein happens throughout the nervous system of the gut, mostly in the salivary glands and the esophagus, less so in the stomach and small intestine, and even less in the colon and rectum, along the many endings of the vagus nerve, a big nerve that connects the gut to the brain.

Deposition of α-synuclein in saliva-producing glands underneath the jaw is only seen in patients with PD and not in healthy adults. If detected, it may serve as a marker of the disease. The nervous system of the gut is affected early in PD. Since they show abnormal α-synuclein deposits, the vagus nerve endings might provide a conduit for abnormal forms of α-synuclein to access the brain.

Misfolding of α-synuclein enables its spread from cell to cell and formation of fibrils and eventually Lewy bodies. Increasing evidence suggests that PD may arise in the gut. Clinically, gastrointestinal symptoms often appear in patients before other neurological signs and aggregates of α-synuclein have been found in enteric nerves of PD patients. Experimentally, abnormal forms of α-synuclein appear in enteric nerves before they appear in the brain and injection of abnormal α-synuclein into the wall of the intestine spreads to the vagus nerve. Ingested toxins and alterations in gut microbiota can induce α-synuclein aggregation and PD. Sensory cells of the gut known as enteroendocrine cells (EECs) contain α-synuclein and synapse with enteric nerves, thus providing a connection from the gut to the brain. It is possible that abnormal α-synuclein first develops in EECs and spreads to the nervous system [12].

The lining of the appendix may also be a site of entry of α-synuclein because this region has the highest density of α-synuclein in the enteric nervous system. Alternatively, increased collections of α-synuclein in the gut and colon might cause breakdown and infiltration of the central nervous system by soluble smaller particles or strains of α-synuclein that climb up into the brain through the vagus nerve potentially with harmful effects. This condition is called synucleinopathy, and it is associated with symptoms along the entire gastrointestinal tract and the central nervous system [13].

It is important to note here that, although there is clear evidence of synucleinopathy in various regions of the gut in PD, it may not be causing gut symptoms. Further research is needed to allow the development of specific therapies for the gut symptoms in this disease.

Evidence For and Against "Gut-First" Instead of "Brain-First" Involvement in PD

Epidemiological Studies

Gastrointestinal symptoms such as constipation, gastroesophageal reflux, and gastroparesis can precede PD motor symptoms by several decades [12]. Large studies have suggested that truncal vagotomy is associated with a substantially reduced risk of developing PD [14]. Non-motor features, such as gastrointestinal dysfunction, could arise earlier than motor features, because the peripheral and enteric nervous system (ENS) cannot compensate for neurodegeneration as the CNS. A large longitudinal study found some reduction in PD risk after truncal vagotomy [15].

In Vivo (Live) Imaging in Patients with Prodromal PD

Multimodal imaging in patients with isolated REM sleep behavior disorder, a recognized prodromal manifestation of PD, found marked peripheral autonomic nervous system involvement, despite relative sparing of the nigrostriatal dopamine system, but the same study examined markers of degeneration without directly measuring pathology [16].

Neuropathological Studies

Autopsy studies have shown synucleinopathy in the ENS in almost all individuals with PD. Such α-synuclein deposition has been found in biopsies from the esophagus, stomach, small intestine, and colon up to 20 years before PD diagnosis. However, other biopsy-based studies detected α-synuclein accumulation in the gastrointestinal tract in neurologically healthy individuals without notable neuronal loss (Fig. 1.4).

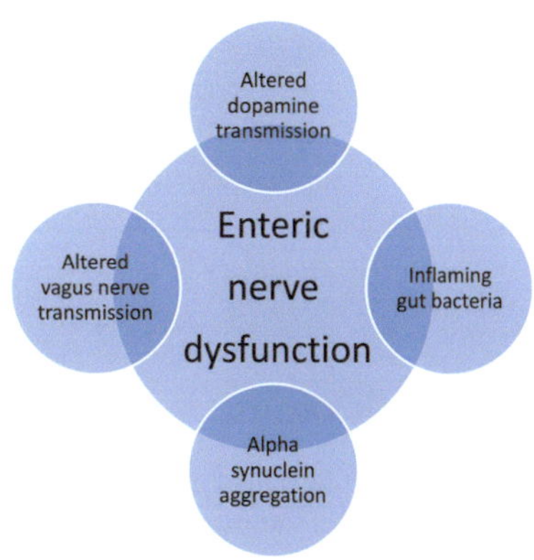

Fig. 1.4 Biologic alterations associated with enteric nerve dysfunction in PD

Gastrointestinal Diseases in PD

Broadly (see diagram), gut PD may result in alterations of function of one or multiple gut organs (i.e., dysphagia, gastroparesis), alterations of both the structure and function (i.e., constipation with associated rectal ulcer or volvulus), or alterations of structure (i.e., *H. pylori* infection-related peptic ulcer with stomach bleeding or obstruction). Any of these alterations may individually or collectively cause symptoms and require differential management (Fig. 1.5).

Functional Disorders in Gut PD

Functional disorders in gut PD can be differentiated in three major categories (see diagram) that may exist in isolation or overlap with each other and require differential therapy (Fig. 1.6).

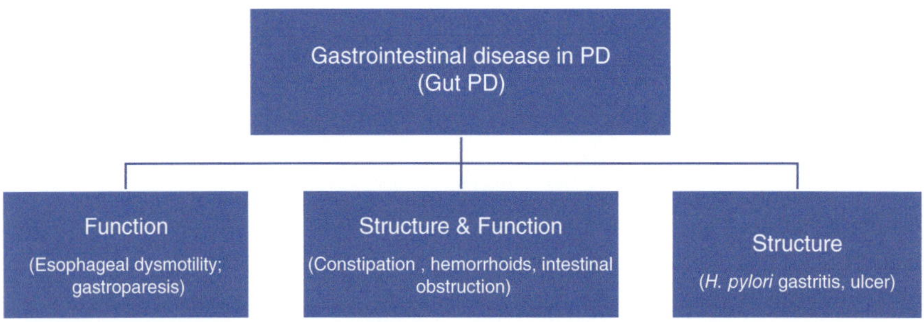

Fig. 1.5 Gastrointestinal disease in PD (Gut PD) may reflect abnormalities of structure, function, or both

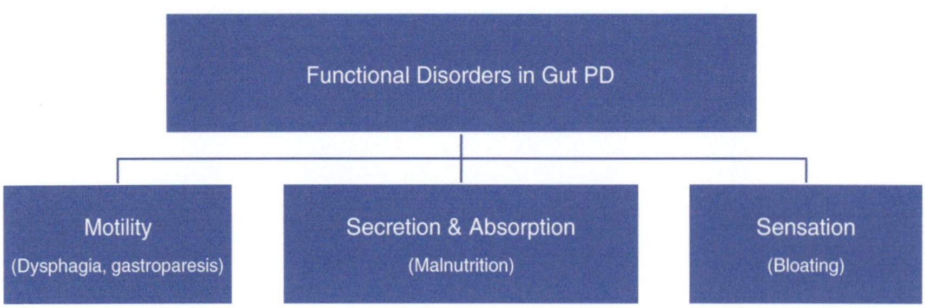

Fig. 1.6 Functional disorders in Gut PD may reflect abnormal motility, secretion and/or absorption, abnormal sensation, or their combination

Structural Disorders in Gut PD

Structural disorders in gut PD can be differentiated in three major categories (see diagram) that may exist in isolation or overlap with each other and require differential therapy. Since most patients with PD are older than 45 years, gastrointestinal cancer needs to be always considered. Structural disorders may arise from inflammation (i.e., peptic ulcer, colitis) and infection (i.e., *H. pylori*, SIBO) or result from poor blood supply (ischemia) (i.e., volvulus, solitary rectal ulcer syndrome) (Fig. 1.7).

Parkinson's Disease (PD) and the Gut (Gut PD)

Over the past decade, we have learned a lot about the importance of gastrointestinal dysfunction in PD. Both the nervous system and the muscles of the gut are frequently affected, causing various symptoms and interfering with the absorption and effectiveness of antiparkinsonian drugs. More importantly, the nervous system of the gut, through the propagation from the gut to the brain of a protein called α-synuclein, may worsen over time the degeneration of the central nervous system.

In this book, we review our present understanding of the gastrointestinal involvement in PD. Many experiments suggest that the gut might be the site where the disease starts. Deposition of α-synuclein along the gut may even serve as a marker of the disease.

We then discuss how the gut dysfunction causes symptoms from the top (mouth) to the bottom (anal canal), and because of their frequency and severity, they contribute to disease progression and disability.

We also discuss the negative effects of small intestinal bacterial overgrowth (SIBO) and of a frequently silent stomach infection by *Helicobacter pylori* infection on the absorption of antiparkinsonian drugs, in turn causing motor fluctuations and inadequate treatment response.

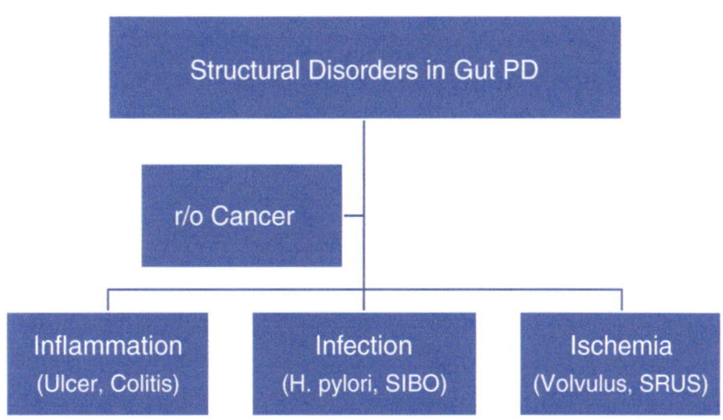

Fig. 1.7 When assessing structure in Gut PD, ruling cancer is a pivotal early step, followed by an evaluation for inflammation, infection, or ischemia

Because of the significant role of the gastrointestinal system in PD, we believe that a full assessment and treatment of gut dysfunction in patients with the disease is essential. We finally present a practical approach for the assessment of the gastrointestinal system in patients with PD and provide evidence-based yet practical recommendations for treatment.

The extrinsic and autonomic nervous system controls the motor, sensory, and secretory functions of the gut through the enteric nervous system (ENS). Gastrointestinal symptoms, or gut manifestations of PD (gut PD), such as sialorrhea, difficulty swallowing, nausea and slow gastric emptying, intestinal pseudo-obstruction, constipation, diarrhea, and fecal incontinence, are quite common in PD. Such symptoms may severely impact the quality of life of these patients, at times as severely as their neurologic disease. Not only these gut manifestations are not readily recognized, but they are frequently inadequately treated.

The diagnosis of gut PD involves not only the identification and assessment of the neurologic disease and its distribution but also the evaluation of diffuse or segmental gut dysfunction, typically using noninvasive imaging, transit measurements, or intraluminal measurements of sensation, pressure activity, and motor coordination. Apart from treatment of the underlying PD, management focuses on restoration of normal hydration and nutrition and pharmacologic treatment of the gut neuromuscular disorder [17].

By outlining the issues from the clinical, diagnostic, and therapeutic perspectives in a clear and well-illustrated fashion, this book will help patients with gut PD, their family, and caretakers to understand the disease and its gastrointestinal impact better. In turn, such understanding will improve the dialogue and communication exchange with their movement disorder specialists (neurologists) and gastroenterologists, as well as other healthcare providers who may—at present—not be in a position for guidance.

There is minimal knowledge–and at times misinformation–on the gastrointestinal dysfunction in this disease among primary neurologists and, surprisingly, even among gastroenterologists who, when asked, they may perform a clinical and endoscopic evaluation by endoscopy and/or colonoscopy and do not fine-tune what is needed, such as issues with underlying dysphagia, gastroparesis, or constipation, to list a few, all of which affecting quality of life above and beyond PD and at times to a similar degree. The impact of associated cognitive decline, resulting malnutrition, and inanition with loss of muscle mass and strength (sarcopenia) in such patients is enormous.

The book will certainly be useful to the parties described above but also to the nonspecialists (primary care, internal medicine, family practice) and advanced practice providers (nurse practitioners and physician assistants) caring for such patients. It will also serve as a primer of gut PD to physical therapists, occupational therapists, nutritionists, emergency department personnel, and general and orthopedic surgeons and as a general guide to the neuro-gastroenterological issues that patients with gut PD suffer from while they are under their care. More importantly, the book will be a guide for sophisticated patients (who wish to synchronize and coordinate their care among health providers and their caretakers, such as spouses or other family members or assistants). Finally, the book will serve as a focal point for exchange of scientific and practical knowledge among PD community groups where, at present, patients and families all over the USA get together to discuss new progress and options for the disease.

Ideally, although not presently generally available, care for patients with gut PD should incorporate and utilize the expertise of many practitioners in tertiary referral centers in the setting of well-structured centers for Parkinson's disease (CPDs) where multidisciplinary resources and experienced personnel are coordinating their services.

Factors in Gut PD

The baseline structural integrity of the gut, from the mouth to the anus, needs to be evaluated in all patients with gut PD, since any abnormalities found could contribute to symptoms and predispose patients to further functional abnormalities. For example, pyloric stenosis due to *H. pylori*-induced peptic ulcer disease may prolong gastric emptying time, simulating gastroparesis, and may respond to pyloric balloon dilation [18]. Underlying diverticulosis with episodic colonic wall edema, muscle thickening, and diverticulitis may worsen bloating, abdominal pain, and constipation. Ongoing neurodegeneration in gut PD eventually may cause luminal dilation and stasis, leading to small bowel bacterial overgrowth (SIBO). In turn, gut microbiome composition alterations affect neurodegeneration through neuroinflammation and may contribute to the pathogenesis and clinical phenotypes (tremor vs. non-tremor) and severity of PD, all driven by host cytokine responses [19]. Additional pill-induced esophageal injury may contribute to episodes of food impaction [20]. Colonic volvulus and Ogilvie's syndrome may also result from progressive chronic colonic dilation or acute pseudo-obstruction [21, 22] (Fig. 1.8).

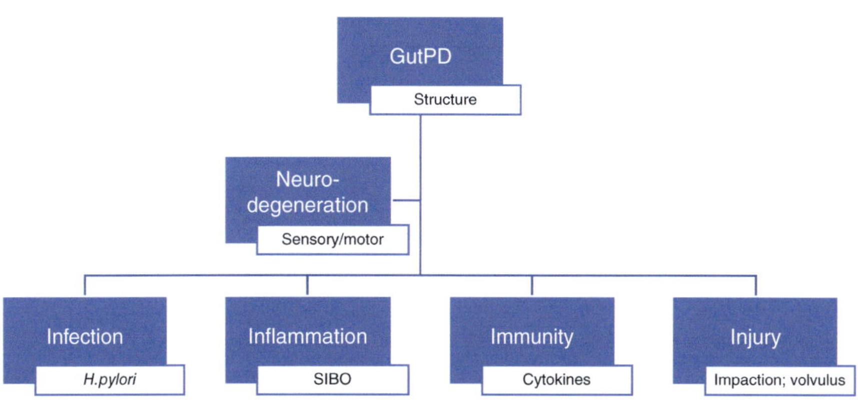

Fig. 1.8 Factors contributing to Gut PD may include nerve degeneration, infection, inflammation, altered immunity, and injury

Gut PD Neuroanatomy and Physiology

The enteric nervous system (ENS), like the brain, contains hundreds of neurons and glial cells, forming complex networks that populate the entire length of the gut wall, from the mouth to the anus and from its serosal to the mucosal layers. Neurons express acetylcholine, nitric oxide synthase, catecholamines, GABA, and various neuropeptides. The ENS varies significantly along the gut axis, serving distinct functions for each individual segment. Different neuronal subtypes and glial cells form neuronal circuits, in turn allowing an autonomous regulation of gut motility and secretion [23].

Although the gut works independently, there is a varying influence of the central nervous system (CNS), mostly through the vagus nerve, forming the brain–gut axis (BGA) that consists of two main routes, one based on humoral factors, such as cytokines, hormones, and gut microbiome metabolites, and another one the vagus nerve that contains thousands of bidirectional fibers [24]. While the afferent fibers deliver information from the gut, the efferent fibers provide parasympathetic motor stimuli that originate in two brainstem nuclei that contribute to gastrointestinal motility [25]. Further influences are exerted by sympathetic fibers from prevertebral ganglia that connect the gut with the spinal cord.

Although to a degree the CNS influences the esophagus and stomach [26], the small and large intestinal smooth muscle activity and their local blood flow, secretion, and absorption are mainly controlled by autonomous reflex circuits, which, if affected, could possibly explain the "top-down" gastrointestinal symptoms in PD. Based on the dual-hit hypothesis [27], there will also be a "bottom-up" process, possibly initiated by local inflammation and a compromised mucosal barrier, allowing gut contents (i.e., bacterial metabolites) to enter the gut wall. In PD patients, the mucosal barrier is compromised and the microbiome is disturbed (dysbiosis). Special nerve circuits monitor the microbiome, or its metabolites lead to modification of the innervation [28]. Overall, the interplay between the BGA, the gut mucosal barrier, and its microbiome could serve as therapeutic targets in PD.

Parkinson's Disease and the Gut

Gastrointestinal (GI) dysfunction is one of the most common non-motor symptoms in PD, occurring in about 60–80% of patients. These include constipation, evacuation disorders, dysphagia, gastroparesis, and small intestinal bacterial overgrowth. Indeed, the GI tract is thought to have an important role in the pathophysiology of PD.

If one dissects the brain of patients with PD, they will find the Lewy bodies or Lewy neurites, characterized by abnormal accumulation of a protein, called α-synuclein. Similarly, abnormal accumulation of α-synuclein happens throughout the nervous system of the gut, mostly in the salivary glands and the esophagus, less so in the stomach and small intestine, and even less in the colon and rectum, along the many endings of the vagus nerve, a big nerve that connects the gut to the brain.

Deposition of α-synuclein in saliva-producing glands underneath the jaw is only seen in patients with PD and not in healthy adults. If detected, it may serve as a marker of the disease. The nervous system of the gut is affected early in PD. Since they show abnormal α-synuclein deposits, the vagus nerve endings might provide a conduit for abnormal forms of α-synuclein to access the brain.

The lining of the appendix may also be a site of entry of α-synuclein because this region has the highest density of α-synuclein in the enteric nervous system. Alternatively, increased collections of α-synuclein in the gut and colon might cause breakdown and infiltration of the central nervous system by soluble smaller particles or strains of α-synuclein that climb up into the brain through the vagus nerve potentially with harmful effects. This condition is called synucleinopathy, and it is associated with symptoms along the entire gastrointestinal tract and the central nervous system.

It is important to note here that, although there is clear evidence of synucleinopathy in various regions of the gut in PD, it may not be causing gut symptoms. Further research is needed to allow the development of specific therapies for the gut symptoms in this disease, as discussed in this book.

Factors Affecting Gut PD Severity

Patients with PD frequently require changes in the timing of administration of their levodopa–carbidopa, altering of the dose, switching to alternative or adjunctive agent, or even other interventions (i.e., intrajejunal levodopa administration) for their motor disease [29]. The management of gut PD remains challenging to neurologists (particularly when symptoms are a result of anti-Parkinson's medications) and gastroenterologists. Unfortunately, there are limited data on the options available, if prior therapies fail. Clinicians need to use their best judgment based on a thorough assessment of the patients' history and objectively demonstrable pathophysiological abnormalities. Since they address various targets, combination therapies, including acid suppressants, promotility agents, and laxatives, may play a role in restoring quality of life and minimizing symptoms. Faced with refractory gut PD, the physician needs to consider the effects of disease on symptoms and quality of life, the burden it imposes to the entire digestive system, both structurally and functionally, and the overall course of the disease and its complications [30] (Fig. 1.9).

Optimizing Patient-Centered Care in Gut PD

The clinical presentation of gut PD is highly heterogeneous and ranges from silent functional disturbance of propulsion, asymptomatic chronic infection by *H. pylori,* to dysphagia, nausea, or chronic constipation. This multitude of covert abnormalities and symptoms—that vary in sensitivity and specificity as well as frequency and severity—matched with their underlying multifactorial nature, poses significant challenges in long-term management and prevention of complications. Therefore,

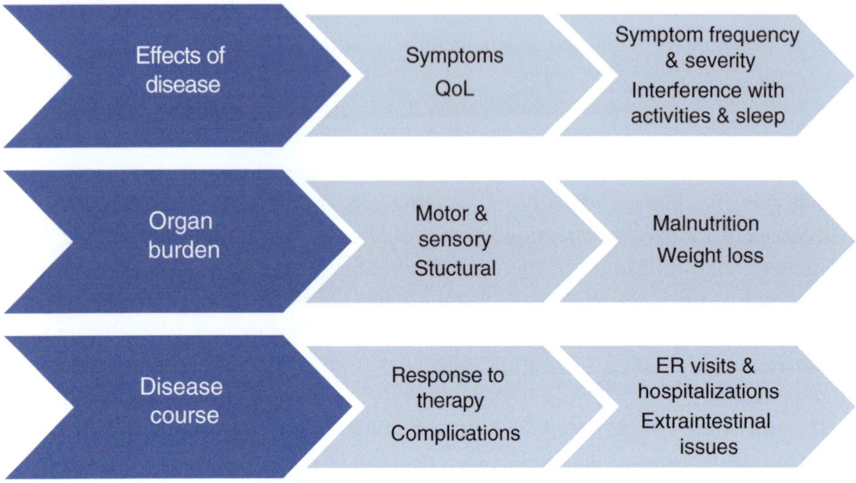

Fig. 1.9 In patients with Gut PD, clinicians should always assess -and periodically reassess- the effects of disease, its burden to various organs, and the disease course

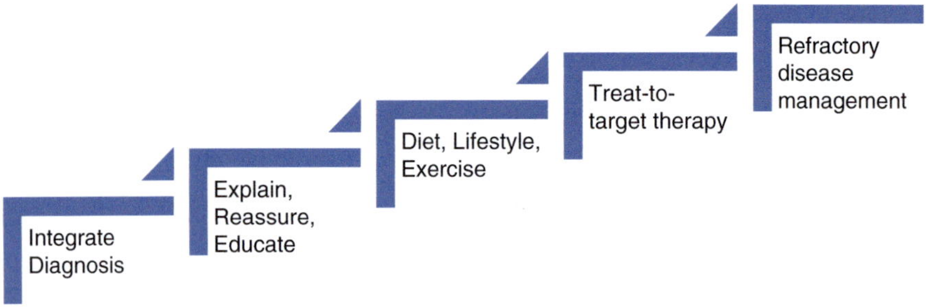

Fig. 1.10 Elements to consider for an optimal, patient-centered, individual approach to Gut PD

to optimize a patient-centered, individualized approach, clinicians need to consider several important elements (Fig. 1.10).

The following diagram broadly outlines what happens in gut PD structurally and functionally, and what steps need to be implemented by the dedicated gastroenterologist in conjunction with movement disorders specialist. Upon establishment of the diagnosis of gut PD, assessment of its severity and risk for complications will assist in defining the need for pharmacologic, endoscopic, or surgical intervention. Disease monitoring is imperative given the progressive nature of gut PD (Fig. 1.11).

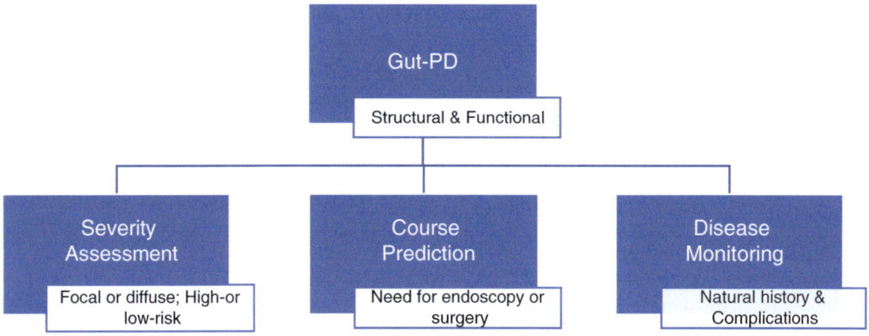

Fig. 1.11 Steps to be implemented by the dedicated gastroenterologist in conjunction with the movement disorders specialist. Disease monitoring over time is essential

References

1. Tanner CM, Ostrem JL. Parkinson's disease. N Engl J Med. 2024;391(5):442–52. https://doi.org/10.1056/NEJMra2401857.
2. Yang W, Hamilton JL, Kopil C, Beck JC, Tanner CM, Albin RL, Ray Dorsey E, Dahodwala N, Cintina I, Hogan P, Thompson T. Current and projected future economic burden of Parkinson's disease in the U.S. NPJ Parkinsons Dis. 2020;6:15.
3. Marras C, Beck JC, Bower JH, Roberts E, Ritz B, Ross GW, Abbott RD, Savica R, Van Den Eeden SK, Willis AW, Tanner CM, Parkinson's Foundation P4 Group. Prevalence of Parkinson's disease across North America. NPJ Parkinsons Dis. 2018;4:21. https://doi.org/10.1038/s41531-018-0058-0. PMID: 30003140; PMCID: PMC6039505.
4. Savica R, Grossardt BR, Bower JH, Ahlskog JE, Boeve BF, Graff-Radford J, Rocca WA, Mielke MM. Survival and causes of death among people with clinically diagnosed synucleinopathies with Parkinsonism: a population-based study. JAMA Neurol. 2017;74(7):839–46. https://doi.org/10.1001/jamaneurol.2017.0603. PMID: 28505261; PMCID: PMC5647647.
5. Fairfoul G, McGuire LI, Pal S, Ironside JW, Neumann J, Christie S, Joachim C, Esiri M, Evetts SG, Rolinski M, Baig F, Ruffmann C, Wade-Martins R, Hu MT, Parkkinen L, Green AJ. Alpha-synuclein RT-QuIC in the CSF of patients with alpha-synucleinopathies. Ann Clin Transl Neurol. 2016;3(10):812–8. https://doi.org/10.1002/acn3.338. PMID: 27752516; PMCID: PMC5048391.
6. Simuni T, Chahine LM, Poston K, Brumm M, Buracchio T, Campbell M, Chowdhury S, Coffey C, Concha-Marambio L, Dam T, DiBiaso P, Foroud T, Frasier M, Gochanour C, Jennings D, Kieburtz K, Kopil CM, Merchant K, Mollenhauer B, Montine T, Nudelman K, Pagano G, Seibyl J, Sherer T, Singleton A, Stephenson D, Stern M, Soto C, Tanner CM, Tolosa E, Weintraub D, Xiao Y, Siderowf A, Dunn B, Marek K. A biological definition of neuronal α-synuclein disease: towards an integrated staging system for research. Lancet Neurol. 2024;23(2):178–90. https://doi.org/10.1016/S1474-4422(23)00405-2.
7. Savica R, Grossardt BR, Bower JH, Ahlskog JE, Rocca WA. Incidence and pathology of synucleinopathies and tauopathies related to Parkinsonism. JAMA Neurol. 2013;70(7):859–66. https://doi.org/10.1001/jamaneurol.2013.114. PMID: 23689920; PMCID: PMC3707980.
8. Kaufmann H, Norcliffe-Kaufmann L, Palma JA, Biaggioni I, Low PA, Singer W, Goldstein DS, Peltier AC, Shibao CA, Gibbons CH, Freeman R, Robertson D, Autonomic Disorders Consortium. Natural history of pure autonomic failure: a United States prospective cohort.

Ann Neurol. 2017;81(2):287–97. https://doi.org/10.1002/ana.24877. PMID: 28093795; PMCID: PMC5323269.
9. Adler CH, Beach TG, Zhang N, Shill HA, Driver-Dunckley E, Mehta SH, Atri A, Caviness JN, Serrano G, Shprecher DR, Sue LI, Belden CM. Clinical diagnostic accuracy of early/advanced Parkinson disease: an updated clinicopathologic study. Neurol Clin Pract. 2021;11(4):e414–21. https://doi.org/10.1212/CPJ.0000000000001016. PMID: 34484939; PMCID: PMC8382425.
10. Gibbons CH, Levine T, Adler C, Bellaire B, Wang N, Stohl J, Agarwal P, Aldridge GM, Barboi A, Evidente VGH, Galasko D, Geschwind MD, Gonzalez-Duarte A, Gil R, Gudesblatt M, Isaacson SH, Kaufmann H, Khemani P, Kumar R, Lamotte G, Liu AJ, McFarland NR, Miglis M, Reynolds A, Sahagian GA, Saint-Hillaire MH, Schwartzbard JB, Singer W, Soileau MJ, Vernino S, Yerstein O, Freeman R. Skin biopsy detection of phosphorylated α-synuclein in patients with synucleinopathies. JAMA. 2024;331(15):1298–306. https://doi.org/10.1001/jama.2024.0792. PMID: 38506839; PMCID: PMC10955354.
11. Morris HR, Spillantini MG, Sue CM, Williams-Gray CH. The pathogenesis of Parkinson's disease. Lancet. 2024;403(10423):293–304. https://doi.org/10.1016/S0140-6736(23)01478-2.
12. Liddle RA. Parkinson's disease from the gut. Brain Res. 2018;1693(Pt B):201–6. https://doi.org/10.1016/j.brainres.2018.01.010. Epub 2018 Jan 31. PMID: 29360467; PMCID: PMC6003841.
13. Tansey MG, Wallings RL, Houser MC, Herrick MK, Keating CE, Joers V. Inflammation and immune dysfunction in Parkinson disease. Nat Rev Immunol. 2022;22(11):657–73. https://doi.org/10.1038/s41577-022-00684-6. Epub 2022 Mar 4. PMID: 35246670; PMCID: PMC8895080.
14. Liu B, Fang F, Pedersen NL, Tillander A, Ludvigsson JF, Ekbom A, Svenningsson P, Chen H, Wirdefeldt K. Vagotomy and Parkinson disease: a Swedish register-based matched-cohort study. Neurology. 2017;88(21):1996–2002. https://doi.org/10.1212/WNL.0000000000003961. Epub 2017 Apr 26. PMID: 28446653; PMCID: PMC5440238.
15. Konings B, Villatoro L, Van den Eynde J, Barahona G, Burns R, McKnight M, Hui K, Yenokyan G, Tack J, Pasricha PJ. Gastrointestinal syndromes preceding a diagnosis of Parkinson's disease: testing Braak's hypothesis using a nationwide database for comparison with Alzheimer's disease and cerebrovascular diseases. Gut. 2023;72(11):2103–11. https://doi.org/10.1136/gutjnl-2023-329685. Epub 2023 Aug 24.
16. Horsager J, Andersen KB, Knudsen K, Skjærbæk C, Fedorova TD, Okkels N, Schaeffer E, Bonkat SK, Geday J, Otto M, Sommerauer M, Danielsen EH, Bech E, Kraft J, Munk OL, Hansen SD, Pavese N, Göder R, Brooks DJ, Berg D, Borghammer P. Brain-first versus body-first Parkinson's disease: a multimodal imaging case-control study. Brain. 2020;143(10):3077–88. https://doi.org/10.1093/brain/awaa238.
17. Camilleri M. Gastrointestinal motility disorders in neurologic disease. J Clin Invest. 2021;131(4):e143771. https://doi.org/10.1172/JCI143771. PMID: 33586685; PMCID: PMC7880310.
18. Artifon EL, Sakai P, Hondo FY, Lopasso FP, Ishioka S, Gama-Rodrigues JJ. An evaluation of gastric scintigraphy pre- and postpyloroduodenal peptic stenosis dilation. Surg Endosc. 2006;20(2):243–8. https://doi.org/10.1007/s00464-005-0129-2. Epub 2005 Dec 28.
19. Lin CH, Chen CC, Chiang HL, Liou JM, Chang CM, Lu TP, Chuang EY, Tai YC, Cheng C, Lin HY, Wu MS. Altered gut microbiota and inflammatory cytokine responses in patients with Parkinson's disease. J Neuroinflamm. 2019;16(1):129. https://doi.org/10.1186/s12974-019-1528-y. PMID: 31248424; PMCID: PMC6598278.
20. Saleem F, Sharma A. Drug-induced esophagitis. In: StatPearls [Internet]. Treasure Island (FL): StatPearls Publishing; 2024.
21. Tateno F, Sakakibara R, Aiba Y, Ogata T, Katsumata M, Matsuoka Y. Recurrent sigmoid volvulus in a patient with Parkinson's disease. Clin Auton Res. 2020;30(3):283–5. https://doi.org/10.1007/s10286-019-00658-0. Epub 2019 Dec 14. PMID: 31838591.
22. Marinella MA. Acute colonic pseudo-obstruction complicated by cecal perforation in a patient with Parkinson's disease. S Med J. 1997;90(10):1023–6. https://doi.org/10.1097/00007611-199710000-00009. PMID: 9347813.

23. Spencer NJ, Hu H. Enteric nervous system: sensory transduction, neural circuits and gastrointestinal motility. Nat Rev Gastroenterol Hepatol. 2020;17(6):338–51. https://doi.org/10.1038/s41575-020-0271-2. Epub 2020 Mar 9. PMID: 32152479; PMCID: PMC7474470.
24. Beyak MJ, Bulmer D, Jiang W, Keating W, Grundy D. Chapter 25: Extrinsic sensory afferent nerves innervating the gastrointestinal tract. In: Johnson LR, editor. Physiology of the gastrointestinal tract. 4th ed. San Diego: Academic Press; 2006. p. 685–725.
25. Travagli RA, Anselmi L. Vagal neurocircuitry and its influence on gastric motility. Nat Rev Gastroenterol Hepatol. 2016;13(7):389–401. https://doi.org/10.1038/nrgastro.2016.76. Epub 2016 May 25. PMID: 27251213; PMCID: PMC5605144.
26. Levinthal DJ, Strick PL. Multiple areas of the cerebral cortex influence the stomach. Proc Natl Acad Sci U S A. 2020;117(23):13078–83. https://doi.org/10.1073/pnas.2002737117. Epub 2020 May 20. PMID: 32434910; PMCID: PMC7293610.
27. Hawkes CH, Del Tredici K, Braak H. Parkinson's disease: a dual-hit hypothesis. Neuropathol Appl Neurobiol. 2007;33(6):599–614. https://doi.org/10.1111/j.1365-2990.2007.00874.x. Epub 2007 Oct 24. PMID: 17961138; PMCID: PMC7194308.
28. Muller PA, Schneeberger M, Matheis F, Wang P, Kerner Z, Ilanges A, Pellegrino K, Del Mármol J, Castro TBR, Furuichi M, Perkins M, Han W, Rao A, Pickard AJ, Cross JR, Honda K, de Araujo I, Mucida D. Microbiota modulate sympathetic neurons via a gut–brain circuit [Erratum in: Nature. 2020 Sep;585(7823):E2.]. Nature. 2020;583(7816):441–6. https://doi.org/10.1038/s41586-020-2474-7. Epub 2020 Jul 8. PMID: 32641826; PMCID: PMC7367767.
29. Jost WH. A novel treatment option for intrajejunal levodopa administration. Expert Rev Neurother. 2023;23(1):9–13. https://doi.org/10.1080/14737175.2023.2176222. Epub 2023 Feb 10.
30. Warnecke T, Schäfer KH, Claus I, Del Tredici K, Jost WH. Gastrointestinal involvement in Parkinson's disease: pathophysiology, diagnosis, and management. NPJ Parkinsons Dis. 2022;8(1):31. https://doi.org/10.1038/s41531-022-00295-x. PMID: 35332158; PMCID: PMC8948218.

Normal Gut Structure and Function

2

Understanding the Essentials of the Normal Gut Movement and Sensation

Normally, the gastrointestinal sensorimotor system works continuously to process the intake of nutrients, such as fats, proteins, carbohydrates, vitamins, and minerals that supply the body with energy and promote tissue growth and repair. All along the length of the gut, from the mouth to the anal canal, complex and highly coordinated nerve, hormonal, and muscular functions allow for the seamless delivery of 1500–3000 calories per day from the diet while simultaneously absorbing 2 L of fluids taken by mouth and 7 L of saliva, stomach juice, pancreatic fluid and enzymes, bile, and intestinal secretions. The gastrointestinal tract is also an integral way for eliminating waste (stool).

These functions require multiple complex processes working together, including chewing, swallowing, and gastrointestinal muscle propulsion passing the contents through the tract and sweeping away waste and debris, secretion of digestive enzymes, absorption of nutrients, and sensation, such as the pleasant feeling after a meal, or the need to evacuate. All of this is coordinated by the brain–gut axis that includes both the central (CNS) and peripheral enteric nervous systems (ENS) and various specialized cells in the gut wall that coordinate nutrient processing and transport during digestion.

The placement of food in the mouth starts the mastication (chewing) reflexes. In the mouth, mechanical receptors send signals to the brain, which in turn coordinates alternating contractions and relaxations of the jaw muscles. Chewing breaks down food boluses while mixing and lubricating them with saliva and its digestive enzymes. Swallowing (deglutition) starts with the mostly involuntary transport of the food bolus from the pharynx (back of the throat) into the esophagus (the swallowing pipe). The soft palate is pulled upward, closing the posterior nares and preventing food from being refluxed into the nose. The larynx (voice box) moves upward against the epiglottis blocking food from entering the trachea and the lungs, and respiration is transiently inhibited. As the food bolus exits the pharynx, the

G. Triadafilopoulos, *The Gut in Parkinson's Disease*,
https://doi.org/10.1007/978-3-031-77171-2_2

swallowing reflex coordinates upper esophageal sphincter (UES) relaxation, allowing the food bolus to enter in the esophagus. The UES then closes, preventing laryngopharyngeal reflux (LPR). A primary esophageal peristaltic wave is then initiated, resulting in a sequence of contractions propelling the bolus away from the mouth toward the stomach. If the bolus is not cleared, a secondary peristaltic wave, triggered by esophageal distention and mediated by the ENS, arises and repeats, until the bolus is cleared. As the bolus approaches the esophagogastric junction (EGJ), the lower esophageal sphincter (LES) opens. As the food bolus moves from the esophagus into the stomach, distention in the lower esophagus triggers relaxation of the proximal stomach food bolus accommodation, in turn decreasing the stomach pressure and increasing its volume. This is followed by a highly coordinated wave of stomach muscle relaxation which allows the peristaltic contraction to move the food bolus toward the distal stomach and duodenum (proximal intestine).

Contractions of the gastric muscular wall propel the food forward, often against the contracted pylorus (a lower stomach sphincter), decreasing the size of the food bolus particles and mixing them with stomach juices. This produces a semifluid mixture under the command of the gastric pacemaker located in the proximal stomach and composed of special pacesetting cells that fire at the muscles to contract by generating spontaneous electrical slow (three to five cycles per minute, cpm) waves and eventually transmitting these waves to the muscle cells in the rest of the gut to generate intestinal contractions, coming and going in phases and ultimately propulsion downstream.

Depending on the timing after a meal and the consistency (solid or liquid) and composition (protein, carbohydrate, fat) of the food in the stomach, these contractions have variable but synchronized intensity, eventually propelling some of the stomach contents through the pylorus into the proximal small bowel (duodenum) when the particle size has been optimized, followed by closure of the pylorus that forces most remaining stomach contents back for more mixing and particle size reduction. Stimulation of the big nerve of the gut (the vagus nerve) together with local hormone release and local distention of the stomach further increase contractility. Sympathetic nerve stimulation and other hormone release (secretin and gastric inhibitory peptide) eventually result in a decrease in the force of stomach contractions and slow emptying into the duodenum, usually within 3 h, that ultimately allows for proper digestion and absorption in the small intestine. Practically, liquids empty faster than solids, carbohydrates faster than proteins, and proteins faster than fats. Concomitantly, local reflexes may suppress stomach emptying, mostly driven by duodenal distention, low pH (acidity), or altered osmolarity of the intestinal contents regulated by multiple locally produced and active hormones.

The small intestinal motility and secretion handle the timely digestion and absorption of nutrients by mixing contents in the gut lumen with digestive enzymes through distention-triggered segmenting contractions. Under the influence of the gut nervous system, distention triggers regularly spaced concentric contractions along the entire length of the small intestine, causing segmentations typically at 12 cycles per minute (cpm). Locally released chemicals trigger contractions, while others facilitate distal gut wall relaxation. Other hormones that circulate in the bloodstream (thyroxine, insulin) increase small bowel peristalsis, while others (glucagon) decrease it. Eventually, increased distention and exposure of contents in the distal small intestine increase peristalsis and relax the valve that leads into the colon (large intestine). Conversely, increased pressure in the proximal colon decreases small intestinal propulsion and increases contraction of the valve between the small intestine and the colon preventing backward flow.

The main function of the right (proximal) colon is the absorption of water and electrolytes from the lumen, while the left (distal) colon desiccates and stores the feces. Segmental contractions mix the large bowel contents facilitating exposure of feces to the colonic mucosal absorptive surface. Triggered by meals, the reflexes between the stomach and the colon or the stomach and the small bowel, or simply colonic distention or irritation, mass movements propel the contents from the right to the left colon usually one to three times per day.

Defecation (expulsion of stool through the anal sphincter) is both involuntary and voluntary. As the most distal part of the colon (rectum) fills with feces, the recto-sphincteric reflex causes the rectum to contract and the internal anal sphincter to relax, creating an urge to evacuate. However, this urge is thwarted by the tonically contracted external anal sphincter, a muscular ring surrounding the internal sphincter, controlled by the brain (voluntary). When it is opportune to defecate, the abdominal wall muscles voluntarily contract increasing the intra-abdominal pressure, the external anal sphincter relaxes, and the rectal smooth muscle contracts, expelling the feces from the rectum.

The Gut Microbiome

The human gut microbiome is made of thousands of diverse microbial species with overlapping metabolic and immune-regulating functions. Gut microorganisms evolve together with the human host to form a complex relationship where the gut provides a nourishing environment for the microbiota to perform various physiological functions. Altered composition and/or function of the gut microbiota associated with an unhealthy or pathological state has been broadly termed "gut dysbiosis" although precisely what constitutes a normal microbiome (normobiosis) remains elusive. Gut dysbiosis can be characterized by an overgrowth of microbiota that have the potential to cause disease, loss of normal commensal organisms, or loss of diversity, leading to a decline in the functional capacity, and the ability to return to the pre-perturbation state of the gut microbiome and perpetuation of dysbiosis.

Various intrinsic and extrinsic factors alter the gut microbiome, leading to marked interindividual variability in microbial composition. Adaptations of the microbiome to changes in environmental and host factors do not always have harmful consequences, and the dynamics of host–microorganism and microorganism–microorganism interactions are important determinants of overall gut microbial composition and function. Microbial compositions among healthy individuals can be highly divergent yet retain quite similar functional capacities. The spatiotemporal structure of the human microbiome, proteome, and metabolome reflects and determines regional intestinal physiology and may have implications for disease.

Since most gut microorganisms reside in the colon, human gut microbiome studies have focused on sequencing bacterial DNA. The multidimensionality of host–microbiome interactions, aging, and intraindividual variability of the microbiome with changes in lifestyle are important factors to consider when interpreting disease-related alterations in gut microbiome composition and function from stool samples, a noninvasive method to interrogate the luminal lower gut community. Little is known about the distribution of microorganisms, their environment, and their biochemical activity in the small intestine because of reliance on stool samples and limited access to only some regions of the gut using endoscopy in fasting or sedated individuals [1].

Gut microbiome changes during aging. There is high interindividual variability in the gut microbiome in older people and distinct differences between community-dwelling older individuals and those residing in long-term care facilities, who tend to have a less diverse or stable microbiome. Microbiome changes in people in long-term care also correlated with dietary factors (i.e., low fiber intake and diets high in saturated fat and/or sugar), impaired nutritional status, and increased markers of inflammation and frailty. Age-related changes in the gut microbiota can be attributable to aging gut physiology (i.e., decreased gut motility and immunity), changes in lifestyle (i.e., diet and physical activity), and health status (i.e., comorbidities, medications, and frailty). The gut microbiota may modulate inflammatory signals and crosstalk with other organs, which is increasingly recognized to contribute to low-grade inflammation that results from chronic activation of the innate immune system during aging.

Reference

1. Shalon D, Culver RN, Grembi JA, Folz J, Treit PV, Shi H, Rosenberger FA, Dethlefsen L, Meng X, Yaffe E, Aranda-Díaz A, Geyer PE, Mueller-Reif JB, Spencer S, Patterson AD, Triadafilopoulos G, Holmes SP, Mann M, Fiehn O, Relman DA, Huang KC. Profiling the human intestinal environment under physiological conditions. Nature. 2023;617(7961):581–91.

Symptoms in Gut PD

3

All patients with PD with symptoms, suspected gut dysfunction, or inadequate response to dopaminergic drugs should be thoroughly evaluated and treated. Finding factors that could be modified is a crucial first step in management. PD treatment should be orchestrated. As an example, because drooling is common during off-times, treatment of fluctuations should first be addressed. Since many antiparkinsonian drugs can cause or exacerbate gastrointestinal symptoms, switching drugs should be considered. Several health professionals, such as speech therapists, dietitians, neurologists, and gastroenterologists, may need to be engaged further highlighting the importance of non-pharmacological, endoscopic, surgical, and rehabilitative management.

Although the symptoms of gut PD can often be attributed to structural or functional abnormality of a specific organ along the length of the gut, mixed presentations are not uncommon, and coexistent issues need to be kept in mind, particularly when patients do not respond to an empirically prescribed therapy, such as acid suppression. Because there is a possibility of such overlap, radical therapies (such as surgery) need to be avoided, unless the diagnosis is ascertained. Table highlights the symptoms originating from various organs of the gut, but overlap may occur (Fig. 3.1).

G. Triadafilopoulos, *The Gut in Parkinson's Disease*,
https://doi.org/10.1007/978-3-031-77171-2_3

Mouth
-Dental problems
-Drooling
-Xerostomia (dry mouth)
-Hypomimia
-Other mouth functional issues (taste and smell)
Pharynx and esophagus
-Dysphagia
-Odynophagia
-Heartburn and regurgitation
-Belching
Stomach
-Fullness and early satiety
-Nausea and vomiting
-Bloating
-Abdominal pain
Small intestine
-Abdominal bloating
-Abdominal pain
-Weight loss and malnutrition
Colon and anus
-Bloating and abdominal distention
-Abdominal pain
-Diarrhea
-Constipation
-Fecal urgency and incontinence
-Rectal bleeding

Fig. 3.1 Key symptoms originating from various organs in Gut PD. Overlap may occur and needs to be addressed appropriately

Mouth

Dental Problems

Because of their poor dexterity, poor jaw opening, and coordination difficulties, patients with PD brush their teeth and seek dental care less frequently than healthy adults. This leads to increased risk for gum problems, caries, and need for teeth extractions. Increased saliva production (sialorrhea) may change the chemical composition of saliva, while, its opposite, dry mouth (xerostomia), might impair the self-cleaning mechanism of the mouth, causing bad breath, altered taste, and poor food intake. Because patients with PD who have removable dentures might have difficulties manipulating them, dental implants might be preferable and should be considered earlier, rather than later, in the course of their disease to mitigate mastication problems.

Drooling

Drooling means excessive pooling of saliva in the mouth that results from either overproduction of saliva or poor clearance of saliva, caused by associated swallowing difficulties. This leads to an inability in keeping the saliva within the mouth. Depending on how it is assessed, the severity of the disease, and presence of speech and swallowing difficulties, the frequency of drooling in PD ranges from 10 to 80%. Drooling may cause social embarrassment, poor mouth hygiene, bad breath, increased accumulation of mouth germs, difficulty eating and speaking, increased risk of aspiration pneumonia, and a profound negative impact on quality of life. Patients with daily drooling have a tenfold risk of lung infection over the course of 1 year, because food and secretions from the mouth are silently aspirated through the larynx into the lungs. The management of drooling depends on its severity and requires multidisciplinary approach that includes daily oral medications to diminish saliva production, periodic injections of Botox for temporary reduction in saliva production, or a variety of open surgical procedures to remove some salivary glands or disconnect others from the mouth [1]. Treatment of drooling includes behavioral interventions, including changes of head position, use of swallow timers or chewing gum, and medications.

Xerostomia (Dry Mouth)

Hypersalivation and hyposalivation and the subjective experience (viz., xerostomia and drooling) are common in patients with PD. In a large study, the prevalence of self-reported xerostomia ranged from 49 to 77%, and that of self-reported drooling ranged from 5 to 80%. The complexity of these salivary problems necessitates a multidisciplinary approach to avoid mutually counteracting treatments from different healthcare professionals [2].

Hypomimia

It is often referred to as a Parkinson's mask, or facial masking; frequent unintentional mouth opening and stooped body posture with dropped head also impair the ability to maintain saliva within the mouth. Age, disease severity, duration, concomitant use of antidepressants, hallucinations, and low blood pressure upon standing also play a role. Increased drooling during off-times (when the effects of antiparkinsonian drugs wear off and symptoms return) suggests that impaired swallowing might be the cause of drooling. If measured properly, saliva production may be reduced in many patients with PD, probably resulting from lack of dopamine. As a result, dry mouth is seen in more than 60% of patients with PD and might coexist with drooling in 30% of them. We do not know if there is a relationship between drooling and α-synuclein deposition in the submandibular glands in patients with PD.

Other Mouth Functional Issues

Impaired sense of smell is well known to happen in PD, but altered or impaired sense of taste may be present in up to 25% of patients, and it may be associated with impaired cognitive function. A feeling of burning in the mouth without any obvious abnormalities on examination has been noted with levodopa treatment. Dry mouth is often a result of medication use, specifically anticholinergics.

Pharynx and Esophagus

Lump in the Throat (Globus)

Globus is the near constant sensation of a foreign body in the throat. History is key in identifying any possible pathology and in directing further investigations and management. Typically, investigations yield negative results. Treatment of globus mainly centers on patient reassurance and counseling, but it may include reflux management, neuromodulation, or speech therapy in selected cases and treatment of any identified cause [3]. When performed, esophageal manometry is abnormal in 63% of patients, and esophageal pH monitoring identifies underlying acid reflux in approximately 20%. Barium esophagram identifies abnormalities in 24% of patients, and endoscopy identifies abnormalities in 22% [4].

Dysphagia

Dysphagia, or difficulty swallowing, can be seen in >75% of patients with PD. It can originate in the mouth and the pharynx because of sluggishness, poor coordination, and rigidity of the swallowing muscles or, more distally, from the esophagus, the tube-like muscle that carries the food into the stomach. The first kind is called *oropharyngeal*, or *transfer* dysphagia, while the other is called *esophageal* dysphagia. Both types can be subjectively reported, but sometimes they are found only upon testing, predisposing patients to aspiration pneumonia. Aspiration may be found in 50% of patients with PD, and it may be silent in 30% of patients. Dysphagia is typically seen in more advanced PD.

Abnormal swallowing may also be found in patients with PD when tested by video-fluoroscopy, barium swallow, or high-resolution esophageal manometry, and a high index of suspicion is needed, since nocturnal poor clearance and stagnation of oropharyngeal and esophageal secretions may lead to silent aspiration pneumonia [5].

There are several factors contributing to dysphagia, making its management challenging [6]. Regardless of their presentation, PD patients may have either oropharyngeal (OD) or esophageal dysphagia (ED) or both [7]. Dysphagia can be a large detriment to the quality of life and likely contributes to the weight loss commonly seen in PD patients. It also increases the risk of aspiration events, which may occur silently in 15–33% of patients [8].

The diagram below outlines the steps to be taken when evaluating a patient with PD who presents with dysphagia. It is always essential to exclude a structural abnormality, such as a benign esophageal stricture or cancer, by endoscopy, before proceeding with other (functional) studies. Radiologic assessment of esophageal transit using fluoroscopic imaging and barium radiography [9], and the assessment of esophageal pressures, peristaltic coordination, and esophagogastric junction (EGJ) relaxation by high-resolution esophageal manometry (HREM) and esophageal pressure topography (EPT) are important clinical tools for the diagnosis of dysphagia and other esophageal symptoms (i.e., heartburn or regurgitation) in such patients [10]. HREM provides greater resolution, minimizes artifacts, and generates three-dimensional topographical plots of intraluminal pressure profiles, increasing the diagnostic accuracy of esophageal dysfunction and clearance in patients with PD [11].

There is a significant decrease in the contractility of the striated esophagus in PD patients compared to healthy volunteers and patients with oropharyngeal without neuromuscular disease [12]. This is important for two key reasons: First, the striated esophagus is near the trachea, and its dysfunction could place patients at increased risk of aspiration pneumonia, the most common cause of death in PD patients [13]. Second, the impaired contractility likely inhibits proper bolus propulsion leading to retention and reflux into the hypopharynx, increasing the risk of laryngeal penetration. Unlike oropharyngeal dysphagia, which can improve with expiratory muscle strength training, there are presently good therapeutic options to strengthen the contractility esophagus [14] (Fig. 3.2).

Patients with PD and swallowing problems should be considered for assessment of both oropharyngeal and esophageal dysphagia. Video-fluoroscopy is the gold

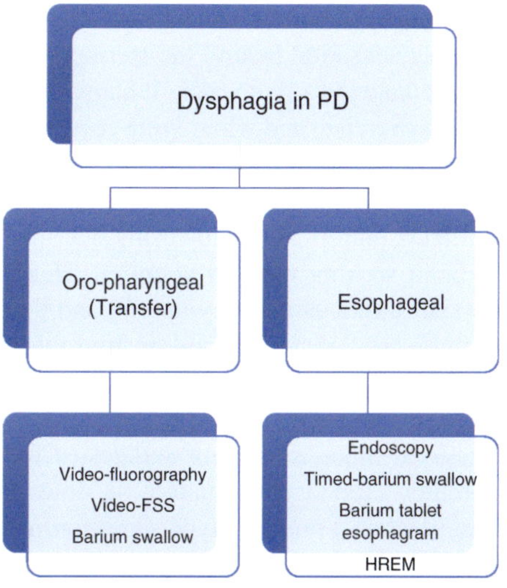

Fig. 3.2 Considerations in patients with Gut PD and dysphagia

standard in the assessment of oropharyngeal dysfunction in patients with PD, while esophageal high-resolution manometry study is the most accurate and sensitive test to identify esophageal dysmotility, which may be found in up to 70% of cases. In patients who are unable to tolerate a manometry study, barium swallow can be used. Endoscopy should always be considered to rule out structural abnormalities. Expiratory chest muscle strength training and video-assisted swallowing therapy may help with oropharyngeal dysphagia; dopaminergic drugs and deep brain stimulation surgery have questionable merit. Drugs have limited value in the treatment of esophageal dysphagia. Instead, depending on the manometry findings, endoscopic Botox injection may help.

Odynophagia (Painful Swallowing)

The most frequent reason for odynophagia in a patient with PD is drug-induced esophagitis, which usually reflects improper administration of pills without adequate water intake and particularly at night. In the proper setting, acute onset of painful swallowing should also raise the suspicion of foreign body impaction, typically from a fish bone, or a foreign body (partial dentures) and will require extraction using endoscopy. Odynophagia should also raise the suspicion of *Candida* or viral esophagitis, both requiring endoscopic confirmation.

Heartburn, Chest Pain, and Regurgitation

Esophageal symptoms, such as heartburn, chest pain, and regurgitation, are common particularly in older adults, and as such they are frequently encountered in patients with gut PD. A careful history should include details on alarm symptoms, such as weight loss, gastrointestinal bleeding, timing of meals, smoking, drug intake, and alcohol consumption. Heartburn is characterized by a discomfort or burning sensation behind the sternum that arises from the pit of the stomach and may radiate toward the neck. It happens intermittently, most frequently after eating, during exercise, and while lying recumbent, and is relieved with drinking water or taking an antacid, but it can occur frequently, interfering with usual activities, including sleep. Because of its association with gastroesophageal reflux disease (GERD), empirical therapy using over-the-counter acid blockers is often done first, prior to seeking medical attention. Heartburn is often confused with indigestion (dyspepsia), a stomach symptom, and the two frequently overlap. Regurgitation is the effortless return of food or fluid into the back of the throat without preceding nausea or retching and at times provoked by bending, belching, or maneuvers that increase intra-abdominal pressure. Patients report a sour or burning fluid in the throat, at times containing undigested food particles or even bile. Chest pain is a common esophageal symptom, at times indistinguishable from cardiac pain; it is described as a pressure-type sensation in the mid chest, radiating to the mid back, arms, or jaws. Esophageal distention or even chemical stimulation by acid will often be perceived as chest pain. Although the most common cause of chest pain is GERD, in patients with PD and underlying esophageal dysmotility, esophageal spasm, drug-induced esophageal injury, or *Candida* esophagitis needs to be considered.

Chest pain, similar in character to angina pectoris, may arise from an esophageal cause, such as GERD, or motility disorder, such as esophageal spasm. Given that most patients with PD are older and at risk for coronary artery disease, further cardiac evaluation is universally advised. The evaluation of esophageal causes of chest pain in patients with PD may involve complementary tests, such as barium swallow, but high-resolution esophageal manometry (HREM) is most often needed to clarify the diagnosis. Chest pain and dysphagia can be triggered by mechanical stimulation, conveyed to the central nervous system via the sensory vagal and spinal afferent nerve fibers. There are three main types of afferent nerve endings, mucosal afferent, intramuscular arrays (IMAs), and intra-ganglionic laminar endings (IGLEs). Calcitonin gene-related peptide (CGRP) is a neurotransmitter expressed in the esophageal mucosa that may play a role in symptom induction [15].

Retrosternal burning, rising from the upper abdomen toward the throat, is part of the spectrum of dyspepsia (indigestion) and is due to reflux of acid and bile into the lower esophagus secondary to excessive relaxation of the lower esophageal sphincter (LES) or its mechanical disruption by a hiatal hernia. In PD, coexisting esophageal dysmotility may further impair acid/bile clearance back into the stomach. Symptoms are typically aggravated by eating large or fatty meals, lying flat (particularly shortly after meals), or stooping. Endoscopic findings do not correlate with symptoms. Often the esophagus appears normal (nonerosive reflux disease, NERD). Other times patients with only mild symptoms have significant reflux esophagitis with inflammation, erosions, or ulcers. Such patients typically have underlying ineffective esophageal motility (IEM) and poor acid clearance. Since many patients with PD are elderly, Barrett's esophagus, that is, the replacement of the distal esophageal squamous mucosa with intestinal-type columnar epithelium, due to chronic acid and bile reflux into the distal esophagus, may be found. A diagnosis of acid and/or bile reflux and esophageal dysmotility is made by high-resolution esophageal manometry (HREM) and ambulatory pH studies [16].

Proton pump inhibitors (PPI) are the best initial therapy for reflux, being superior to H2-receptor antagonists, or antacids alone. Alginates are often useful as "as needed" treatments. Lifestyle alterations, including weight reduction, avoidance of fatty foods, and raising the head of the bed, may help. In some patients with long-standing *Helicobacter pylori*, a gastric antrum-dwelling bacterium may induce acid reflux. Endoscopic techniques for treating GERD include esophageal radiofrequency energy delivery/Stretta procedure and transoral incisionless fundoplication (TIF) and endoscopic full-thickness plication. Other novel techniques include anti-reflux mucosectomy, peroral endoscopic cardiac constriction, endoscopic submucosal dissection, and endoscopic band ligation [17, 18].

In refractory patients with well-validated GERD, particularly those with significant hiatal hernia, surgical treatment is a laparoscopic fundoplication whereby the fundus is wrapped behind the esophagus and the hernia is reduced back into the abdomen. In PD patients with compromised esophageal motility, complications may include dysphagia and bloating due to an inability to belch. Long-term (>20 years) experience with laparoscopic fundoplication carries a success rate of 80.4% among uncomplicated GERD patients and 63.3% among patients with large hiatal hernias [19].

Belching

Eructation (belching or burping) is a common benign—yet distressing—symptom that is frequently encountered in patients with PD, but it is also reported by normal individuals as well as in patients with dyspepsia or gastroesophageal reflux disease (GERD). The belch reflex is a physiological mechanism to expel swallowed air (aerophagia) from the stomach and therefore may be more common after meals. The reflex consists of transient lower esophageal sphincter (LES) relaxation, gas reflux from the stomach to the esophagus to produce a "common cavity" event, upper esophageal sphincter (UES) relaxation, esophago-pharyngeal gas reflux, and restoration of esophageal pressure to baseline pressures through effective esophageal peristalsis [20]. Normally, physiological acid reflux may occur after meals in association with possible leakage of acid as gas is refluxed through the relaxed LES. Belching and acid reflux in normal adults often occur simultaneously. In addition, transient LES relaxations are observed during belching events. Studies with air insufflation have shown that transient LES relaxations are provoked by stomach distention similar to those seen before pathological acid reflux events in patients with GERD [21]. Excessive belching in PD has often been attributed to aerophagia and treated with behavior modification. However, since belching is a manifestation of GERD, the diagnosis of GERD should be considered, and ambulatory pH monitoring performed to assess for symptom association and severity of esophageal acid exposure that would be amenable to medical, surgical, or endoscopic therapy [22]. Further, underlying functional esophageal dysmotility, functional dyspepsia, and gastroparesis in patients with PD may play a role, and they will need further assessment, using HREM and gastric scintigraphy.

Stomach

Early Satiety, Bloating, and Distention

Early satiety is the inability to eat a full meal or feeling full after only a small amount of food. The feeling of early satiety that occurs soon after eating and limits the patient's ability to complete a meal is a significant symptom, since it contributes quickly to weight loss and malnutrition. Although, in the patient with PD, one should consider slow (delayed) stomach emptying (gastroparesis) as a common culprit, an endoscopy is absolutely indicated to first rule out structural abnormalities, such as cancer of the esophagogastric junction or the stomach, or even benign abnormalities, such as pyloric outlet obstruction from peptic ulcer disease. Early occurrence of upper abdominal bloating and distention may also be accompanying early satiety, and they are mostly localized in the upper mid-abdomen, in contrast to bloating and distention that may occur 1–2 h after meals, usually originating from the small intestine or the colon, that are more diffuse in their localization. Frequent eructation (belching) and halitosis (bad breath) frequently accompany delayed stomach emptying and may significantly compromise the patient's quality of life. Proper diagnosis and effective therapy of gastroparesis may mitigate their occurrence. This is most likely due to gastroparesis, a condition in which the stomach is slow to empty and is typically seen in late PD. The patient feels full soon, less that 30 min, after initiating a meal and is unable to complete a normal portion, eventually leading to weight loss, nutrient deficiency, and malnutrition. Since early satiety may also be a sign of serious medical conditions, such as stomach cancer, ulcers, or diabetes, an endoscopy is imperative to rule out structural reasons.

Indigestion

Indigestion (dyspepsia) is a vague term that describes a constellation of symptoms, such as upper abdominal discomfort or bloating, retrosternal or epigastric pain, nausea, loss of appetite, and early satiety after a meal. Epigastric discomfort (dyspepsia) is an ill-defined, aching sensation in the upper mid-abdomen, at times described as hunger pain. If it is related to an underlying duodenal ulcer, it occurs 1–3 h after a meal and is frequently relieved by antacids or food while at times awakening the patient from sleep. The discomfort in patients with gastric ulcer may be precipitated by food, while nausea and weight loss occur more often. Non-ulcer dyspepsia is the condition where, after a careful endoscopy, no gastric or duodenal ulcer and cancer are found. Possible explanations for dyspepsia include acid-, pepsin-, or bile-induced activation of chemical receptors in the proximal small intestine or of altered gastroduodenal motility. Dyspepsia is commonly a result of the use of multiple medications or supplements, and a thorough review is needed before proceeding with any further diagnostic or therapeutic efforts.

Dyspepsia affects up to 40% of the population, and in only 25% of cases is a cause found. Alarm features, such as age over 50, new onset of symptoms, persistent dyspepsia despite empiric treatment, history of peptic ulcers or cancer, NSAID or aspirin use, associated dysphagia, weight loss, or vomiting, call for an upper endoscopy to rule out structural abnormalities, particularly cancer, and *Helicobacter pylori* infection. *H. pylori* infection may induce gastroparesis, resulting in a prolonged transit time for levodopa to reach the small intestine. *H. pylori* may exploit levodopa to facilitate its own growth and proliferation, or it may damage the mucosa, leading to gastrointestinal ulcers and impeding levodopa absorption [23]. Non-ulcer (functional) dyspepsia is a variant of irritable bowel syndrome (IBS). An upper abdominal ultrasound may be justified to exclude gallstones. The optimal treatment for non-ulcer dyspepsia is unknown, but acid suppression is helpful for many. A key consideration in patients with PD is to exclude the presence of gastroparesis that will require gastroenterology consultation and possibly more aggressive therapy [24].

Loss of Appetite (Anorexia)

The etiology of the anorexia in aging adults, particularly those with PD, is multifactorial and includes a combination of decline in smell and taste, reduced central and peripheral drive to eat, delayed gastric emptying, pathological conditions (depression, dementia, medications and iatrogenic interventions, oral health status), and social factors (poverty, loneliness). Anorexia of aging is associated with protein-energy malnutrition, sarcopenia, frailty, functional deterioration, morbidity, and mortality. Since this symptom can lead to dramatic consequences, early identification and effective interventions are needed [25].

Nausea and Vomiting

Nausea is a constellation of unpleasant symptoms including bloating, postprandial fullness, early satiety, and a deep sensation of an impending need to vomit. Nausea may be acute or chronic and may not reflect gastrointestinal disease, since it may result from a systemic, metabolic disease or a central nervous system condition. Acute causes of nausea are usually a result of dietary aberrations (i.e., fatty or spicy foods or alcohol), medications (i.e., nonsteroidal anti-inflammatory drugs, NSAIDs), or acute gastroenteritis. Chronic nausea may reflect gastritis, duodenitis, reflux esophagitis, peptic ulcer, or *Helicobacter pylori* gastritis and is usually caused by medications or metabolic abnormalities. In patients with gut PD, motility disorders such as gastroparesis or chronic idiopathic intestinal pseudo-obstruction should be considered. However, given the older age of patients with gut PD, upper abdominal cancers (i.e., stomach, pancreas) need to be first excluded by endoscopy or abdominal CT.

Vomiting is the forceful ejection of stomach contents (i.e., food, bile), and it may or may not be associated with nausea. Vomiting should be distinguished from regurgitation whereby fluid, or food, is expelled from the esophagus before reaching the stomach. Complete or partial gastric or small intestinal obstruction or even the colon needs to be considered when vomiting is acute or recurrent. The underlying cause of a patient's vomiting is often indicated by the interval between eating and vomiting which gives some indication of the level of a mechanical obstruction. An abdominal CT is the test of choice to initiate the diagnostic evaluation of such patients.

Important mechanical causes for vomiting in gut PD include volvulus of the stomach or the sigmoid colon which present with upper abdominal pain and abdominal distention. Vomiting may also be a feature of other acute conditions, such as acute appendicitis and biliary or renal colic. Chronic slow emptying of the stomach (gastroparesis) in PD is best diagnosed with gastric scintigraphy. Vomiting may also be a feature of intracranial pathology (i.e., space-occupying lesions, cerebrovascular accident, or hydrocephalus) or systemic disease such as sepsis or diabetic ketoacidosis. Complications of vomiting include Mallory–Weiss tear of the esophagus, characterized by vomiting followed by hematemesis, or rupture of the esophagus

(Boerhaave's syndrome), presenting with vomiting followed by severe, acute retrosternal chest pain. Protracted vomiting leads to the development of a hypokalemic alkalosis due to loss of gastric acid.

Gastroparesis, or delayed emptying of food from the stomach after meals, is noted in more than 70% of patients with PD, in both early and advanced diseases, and it may cause no symptoms. When present, such symptoms include nausea, vomiting, early satiety and excessive fullness after meals, bloating, and abdominal distention. Such symptoms are more likely to occur with solid meals, particularly greasy or fatty, and less so with soups or liquids. Gastroparesis is most likely a result of involvement of the gut nervous system, but it can also reflect brain involvement. Abnormal ghrelin function might be involved, since this so-called hunger hormone is normally secreted when the stomach is empty, thus increasing gastrointestinal motility in preparation for ingestion of food. Since levodopa is absorbed in the small intestine, patients with gastroparesis may experience delayed delivery of the drug or complete loss of its benefit due to poor timing of absorption and leading to an unpredictable duration of action.

Abdominal Pain

Epigastric abdominal pain that originates from the stomach typically is located just underneath the xyphoid, at the pit of the stomach, and remains in the upper abdominal location with rare extension in the right and left upper abdominal regions. The pain is variable in severity but related to meals, typically limiting food intake because of its rapid onset after a meal. When this occurs, a stomach ulcer or cancer needs to be considered, and an endoscopy is urgently justified.

Bloating and Abdominal Distention

This is a hard-to-define symptom where the patient may describe their abdomen feeling full (bloating) or bigger than usual, either soon after meals (early satiety) or for hours thereafter. Upper abdominal pain or discomfort, abdominal rumbling, and even excessive belching and passing of gas are also noted frequently. As a symptom complex, it could reflect a stomach, small bowel, or colonic abnormality.

The table provides a simplistic approach to understanding the reasons behind bloating and abdominal distention, based on the symptom onset after a meal and its possible relationship to an underlying organ dysfunction. However, overlap may frequently occur. When bloating occurs soon after meals and is not associated with significant distention but instead with a sensation of early satiety that limits the completion of a meal, one should consider the stomach as the primary source of the problem. PD-related gastroparesis should be explored by performing wireless motility capsule testing, gastric scintigraphy, or electrogastrogram. Bloating that is not clearly related to meals is usually long-lasting and, -if not alleviated with belching or passing gas- is most likely due to small bowel dysmotility, often associated

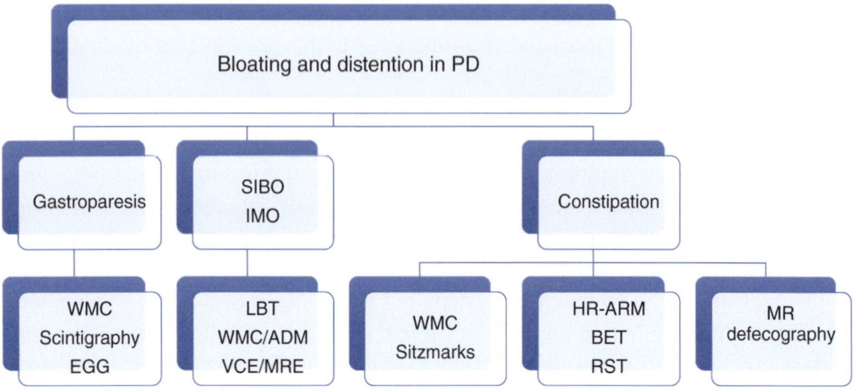

Fig. 3.3 Important considerations in patients with PD presenting with bloating and abdominal distention

with small intestinal bacterial overgrowth (SIBO) or intestinal methanogen overgrowth (IMO). In such cases, lactulose breath testing (LBT), wireless motility capsule (WMC) testing, or antroduodenal manometry (ADM) should be considered. If these studies are not available, alternate options could be the use of video capsule endoscopy (VCE) or MR enterography (MRE) to indirectly assess small bowel transit and rule out structural issues at the level of the small bowel. When bloating and abdominal distention are not readily related to meals, and they transiently improve by passing of stool or gas, a colonic origin should be sought. After excluding a structural obstruction (diverticulosis, stricture, or colon cancer) by colonoscopy, measuring colon transit time using WMC or Sitzmarks testing should be considered, since slow transit constipation, due to generalized colonic inertia, is a possibility. High-resolution anorectal manometry, balloon expulsion testing, and rectal sensory testing assess the sphincter competence and neuromuscular coordination during defecation. Except for lower rates of anal hypertonicity in patients with slow colon transit time, disorders of recto-anal coordination and rectal sensation are commonly seen [26]. MR defecography is distally reserved for patients with defecatory dysfunction, typically elicited by a careful history (Fig. 3.3).

Small Intestine

Abdominal Bloating

Bacteria are the sole producers of intestinal hydrogen, and the LBT is based on bacterial metabolism of ingested lactulose into hydrogen. Normally, the small bowel contains only small numbers of bacteria, and ingested lactulose is metabolized when it reaches the dense population of bacteria in the colon, giving rise to a late peak in breath hydrogen content. In patients with small intestinal bacterial overgrowth (SIBO), the lactulose metabolism by bacteria produces an early rise of breath hydrogen. SIBO, present in up to 70% of patients with gut PD, may cause nonspecific abdominal distention and bloating, irregular bowel evacuation pattern, and foul-smelling stool and is at times associated with worse motor function [27].

Abdominal Pain

Abdominal pain in patients with gut PD may be acute or chronic, focal or generalized, and isolated or occurring together with other symptoms that may help the differential diagnosis. Depending on the patient's disease severity and function, ambulation, or cognitive ability, variable levels of diagnostic approaches may be implemented. In general, common causes of acute (surgical) abdomen need to be first ruled out. Often, abdominal pain is localized with special features (i.e., biliary colic, acute appendicitis). Peptic ulcer, acute peritonitis, and intestinal obstruction may be associated with significant morbidity and mortality. Other times, the pain may be nonspecific and localizing and often functional in nature. In either case, abdominal pain should always be investigated for an underlying structural or functional etiology and properly and expeditiously treated. A thorough history and physical examination should always be undertaken, since patients with PD are often elderly and vulnerable to quick, generalized decline in their general medical state. On abdominal examination, distention with visible peristalsis may be present accompanied by increased bowel sounds and audible borborygmi. In contrast, reduction or paucity of bowel sounds suggests adynamic ileus, or peritonitis. Examination of the hernial orifices is essential to avoid missing a strangulated inguinal, femoral, or umbilical hernia.

In patients confined in chronic nursing facilities, a transfer to a nearby emergency department may be mandatory, since it will allow expeditious imaging and possibly intervention, as needed. A rectal examination is also essential not only to rule out the ballooned empty feel of intestinal obstruction but also to rule out fecal impaction as a cause.

One of the most difficult problems in gut PD is the patient who presents with severe chronic generalized abdominal pains and in whom all clinical, laboratory, and imaging studies are negative. Various structural and functional studies will then need to be implemented.

Weight Loss, Malnutrition, and Frailty

Malnutrition may be seen in up to 25% of patients with PD and is a key factor for the general health and well-being particularly in the elderly, causing functional and mobility limitations, due to weakness and muscle atrophy (sarcopenia), and in turn reducing daily quality of life and overall survival. Aging increases nutritional risk because of diminished organ system reserves. Up to 70% of older adults are malnourished, which in turn is associated with increased mortality. The nutritional needs of older adults depend on multiple factors, most importantly disease-related organ dysfunction, and the person's energy expenditures as well as the ability to access, prepare, ingest, and digest food. The diagnosis requires the combination of at least one **phenotypic** (involuntary weight loss, low body mass index (BMI), or reduced muscle mass) and one **etiologic** criterion, such as reduced food intake or absorption, or underlying inflammation, due to acute disease/injury or chronic disease [28] (Fig. 3.4).

After establishing the diagnosis, a comprehensive assessment of nutritional status should be performed by dietitians or nutritionists, to design and develop individualized care and treatment. Screening involves serial measurements of weight, to calculate the % weight loss, and utilizing screening tools, such as the Mini Nutritional Assessment (MNA). Weight loss is significant if it exceeds 5% in 3 months, or more than 10% in 6 months [29].

Involuntary weight loss may be driven by a variety of factors, such as inadequate dietary intake, loss of appetite, muscle atrophy (sarcopenia), inflammatory effects of disease (cachexia), or a combination of these factors. Medical and psychiatric factors, such as malignancy and depression, play a key role. Many factors may contribute to malnutrition and may include loss of smell and taste, dysphagia, fear of increased off-time, fasting associated with drug administration, drug-induced nausea, and poor appetite all affecting food intake and absorption [30]. Female sex, old age, depression, cognitive decline and apathy, loss of appetite, depression, or apathy may contribute. Further, the enhanced muscle activity or dyskinesia may cause weight loss by increasing energy expenditure. Treatment with dopamine agonists and neurosurgical procedures, such as deep brain stimulation (DBS), may lead to nutritional improvement and weight gain (Fig. 3.5).

Often resistant to nutritional intervention, cachexia is a complex metabolic syndrome associated with underlying illness, characterized by loss of muscle with or without loss of fat mass, anorexia, inflammation, insulin resistance, and increased muscle protein breakdown. Sarcopenia, a syndrome characterized by the loss of muscle mass, strength, and performance without underlying illness, is associated with increased rates of functional impairment, disability, falls, and mortality [31].

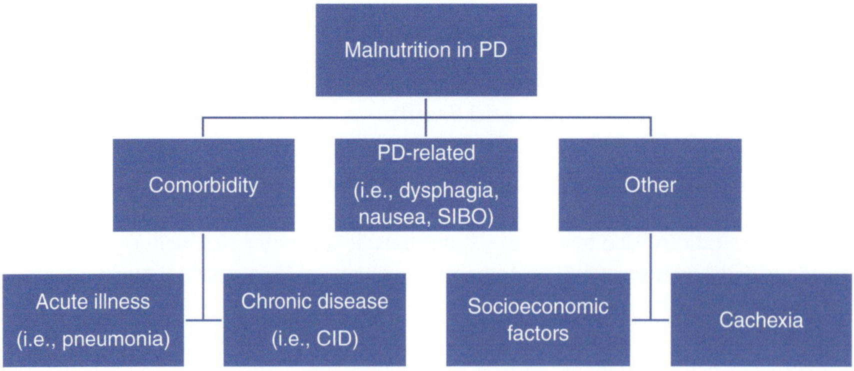

Fig. 3.4 Diagrammatic approach to malnutrition in patients with PD

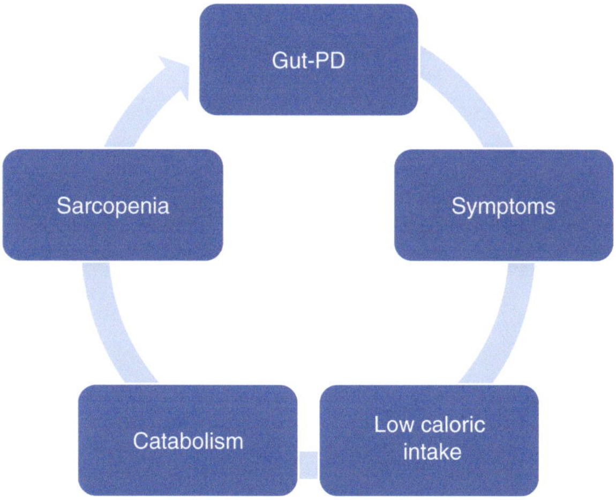

Fig. 3.5 Cycle of events in Gut PD leading to malnutrition, sarcopenia, and cachexia

To evaluate weight loss and malnutrition, one should start by performing a complete history and physical examination, followed by laboratory evaluation for evidence of metabolic or inflammatory disease, such as comprehensive chemistry profile, thyroid-stimulating hormone (TSH), complete blood count (CBC), and C-reactive protein (CRP). A dietician or nutritionist should assess if calorie and protein requirements are met and, together with clinical criteria, grade the severity of malnutrition.

Frailty is a state of diminished physiological reserves and increased vulnerability to disease. Frailty becomes more common with age, particularly in patients with advancing PD. Assessing frailty enables clinicians to predict the outcomes and health risks of health conditions and tailor clinical management. The Fried frailty phenotype delineates a clinical syndrome resulting from altered metabolism coupled with abnormal stress responses [32]. Key features are exhaustion, weakness, slowness, physical inactivity, and weight loss. The presence or absence and degree of frailty are determined by the number of features present: A person is “robust” if none of the features are present, “prefrail” if one or two are present, and “frail” if three to five are present, independent of other coexisting conditions [33]. Frailty assessment should be used to facilitate and optimize care (e.g., exercise, nutritional support, etc.). The biologic mechanisms of frailty are incomplete. Aging is associated with a decline in anabolic hormones and an increase in catabolic hormones (e.g., cortisol) [34]. These hormonal changes inhibit the growth of skeletal muscle and promote its breakdown. Chronic inflammation, which may occur in response to noninfectious triggers such as cellular senescence and mitochondrial dysfunction, inhibits growth factor expression and increases catabolism, thereby contributing to sarcopenia and frailty [35] (Fig. 3.6).

Many instruments are available to measure frailty. These tools rely on patient report (e.g., the FRAIL [Fatigue, Resistance, Ambulation, Illnesses, and Loss of Weight] questionnaire) [36]. Frailty assessment can enable clinicians to predict the outcomes and risks of age-related health conditions and tailor clinical management

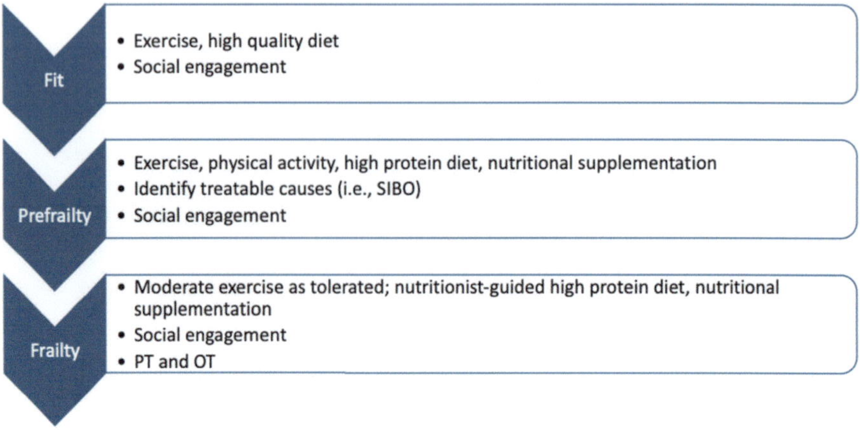

Fig. 3.6 Stages of evolution towards frailty in Gut PD and opportunities for intervention and prevention

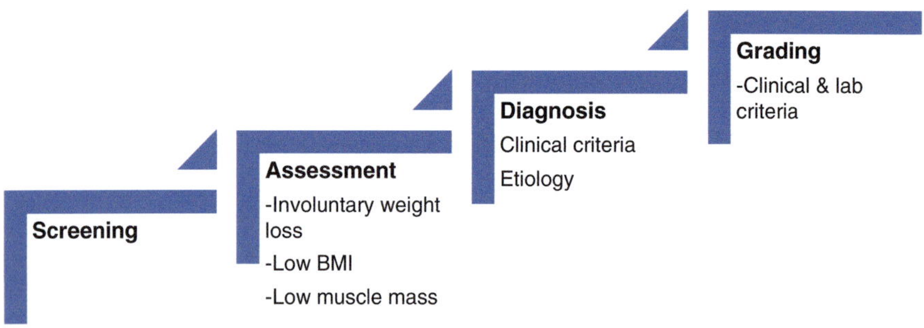

Fig. 3.7 Steps to mitigate progressive malnutrition and frailty in patients with Gut PD

aiming at increasing physiological reserve, building physical resilience, and preventing stressors. Care for persons without frailty should focus on increasing physiological reserve through a healthy lifestyle, management of chronic disease, and preventive care.

In resistant cases where symptom-focused simple modalities and behavioral changes have failed, the use of appetite stimulants may be cautiously considered. Megestrol acetate may lead to weight gain in patients with anorexia and cachexia but is risky in elderly patient with PD. Dronabinol may be useful for anorexia, weight gain, and behavior problems in patients with advanced PD who were refusing food but has significant central nervous system side effects, limiting its use. Mirtazapine, an antidepressant that leads to more weight gain than selective serotonin reuptake inhibitor (SSRI) antidepressants, is commonly used for management of depression and weight loss in patients with PD [37] (Fig. 3.7).

Colon and Anus

Bloating, Belching, Flatulence, and Abdominal Distention

Normally, the volume of gas in the gut (200 mL) is relatively constant, both during fasting and after meals. Abdominal bloating, distention, and excessive passing of flatus are common complaints in patients with PD. This is often due to involuntary poorly coordinated activity of the diaphragmatic and anterior abdominal wall muscles, rather than excessive amounts of intestinal gas. Intermittent belching is commonly caused by swallowing air (aerophagia), typically while eating fast and talking, and eating certain foods such as onions, cabbage, or fatty foods that delay stomach emptying. Excessive bloating may also be a symptom of lactose intolerance, gluten sensitivity, small intestinal bacterial overgrowth (SIBO), or delayed gastric emptying without mechanical obstruction (gastroparesis). The latter two conditions are prevalent in PD and require special consideration. Generalized abdominal bloating is mostly due to irritable bowel syndrome (IBS) and/or diverticulosis and is often associated with altered bowel habits, such as alternating constipation and diarrhea. Some foods contain high amounts of simple, poorly digested carbohydrates which are fermented by colonic bacteria leading to excess gas production and the formation of osmotically active substances. Reducing wheat, onions, broccoli, beans, and insoluble fiber may be useful.

Constipation

Constipation is seen in 80% of patients with PD. It is also one of the earliest non-motor symptoms to occur in PD and is considered a factor that increases the risk of future PD. Importantly, constipation may emerge long (10–15 years) before the onset of the motor symptoms. The risk of development of PD increases at least threefold as the constipation becomes more severe. The underlying mechanisms are unknown, but reduced density or even absence of the gut nerves has been seen in patients with severe chronic constipation. Constipation is also associated with cognitive decline in PD, and its severity can determine the progression to dementia in such patients. Constipation often goes unrecognized during the early stages of PD. As the disease progresses, the severity and frequency of constipation often worsen, causing a significant burden to patients and their caregivers. Early recognition and optimal medical management of constipation not only can improve quality of life but also lessen the chance of emergency visits for fecal disimpaction and hospitalizations for endoscopic or surgical therapy of twisted colon (volvulus) or perforation with peritonitis.

There are several reasons for constipation in patients with PD, such as slow transit constipation because of loss of nerve function in the gut wall or infiltration of Lewy bodies in the vagus nerve that connects the brain to the gut. Increasingly recognized and undertreated is the presence of pelvic floor muscle incoordination (dyssynergia), with paradoxical anal sphincter and puborectalis muscle contraction during defecation, causing poor fecal expulsion (evacuation disorder), with or without impaired rectal sensation due to nerve loss.

In patients with chronic constipation, a thorough history and digital rectal examination are helpful to differentiate colonic transit abnormality from dyssynergic defecation. Other useful studies are the radiopaque marker study for colonic transit measurement, anorectal physiology testing, and MRI defecography. Anorectal physiologic tests, such as high-resolution anorectal manometry (HR-ARM), rectal sensory testing (RST), and balloon expulsion tests (BET), are presently recommended as the initial steps in the evaluation of chronic constipation. Studies examining the HR-ARM profiles across patients with PD have variably identified loss of local reflexes, dyssynergic defecation, hypercontractile anal sphincter, and/or impaired anal squeeze pressures.

Constipation is a common side effect of many drugs used in PD, particularly anticholinergics and dopamine agonists. Nevertheless, delayed colonic transit has been reported in patients with PD, independent of drugs. More frequently, patients might also have poorly coordinated (dyssynergic) bowel evacuation, which is characterized by the inability, or paradoxical increase, of the pelvic floor and anal canal muscles to relax during attempted defecation. Although dysfunction of the sacral and pelvic ganglia is suspected, the underlying reason for such evacuation disorder in PD might involve dysfunction of the effect of dopamine.

Chronic constipation (CC) is defined as stool evacuation frequency of <3/week, usually described as "too infrequent, too little, too hard" usually for more than 3 months. Lack of urge to defecate suggests slow transit constipation (STC). Defecatory straining, a sense of incomplete stool evacuation, and the need to use digital maneuvers (through the anal canal into the rectum or through external pressure through the vagina) suggest evacuation disorder (ED). These two conditions may coexist in many patients with PD (mixed constipation, MC). Obstructive constipation (OC) is a late form of fecal impaction (FI) (hardened stool trapped in the rectum). FI may go unnoticed, typically confused as worsening constipation, or it may manifest as acute intestinal obstruction (bowel blockage), or colonic volvulus (twisting of the colon), causing severe abdominal bloating, distention, and pain, and may lead to death if left untreated. Under these circumstances, an emergency evaluation, clinical examination (which includes a digital rectal examination by a physician), and the use of blood tests, abdominal radiographs, and CT imaging are needed. Depending on the preliminary results, urgent attention by a gastroenterologist to perform a decompressing colonoscopy and release (untwisting) of volvulus or a surgeon to perform abdominal surgery is needed. Many such patients get hospitalized (Fig. 3.8).

There are many causes of constipation in adults, as well as patients with PD, and they always need to be considered. Mostly, constipation is functional and not related to significant underlying disease. In such cases, which can be acute, intermittent, or chronic, no abnormalities are found upon testing. Relief can be spontaneous or result from changes in the diet, fluid intake, or the transient use of laxatives. In patients with chronic or recurrent constipation, secondary causes need to be

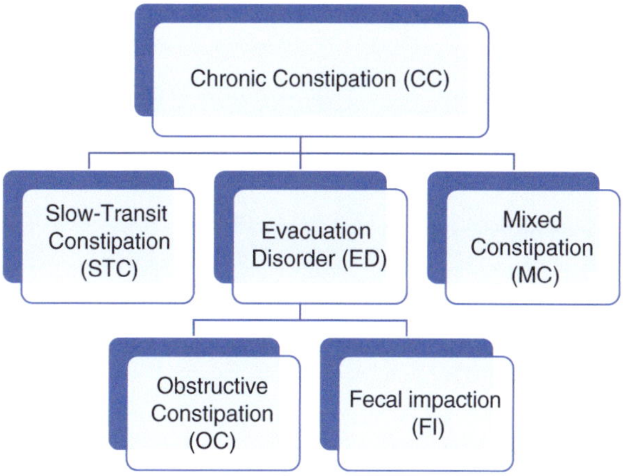

Fig. 3.8 Algorithmic approach to chronic constipation in patients with Gut PD, based on dominant underlying pathophysiology

considered. In patients with PD, constipation may result from drugs (anticholinergics, dopamine, narcotics, etc.), from the disease itself, from other metabolic causes (diabetes, low thyroid function, electrolyte disturbances, etc.), or a combination. Mechanical obstruction by a colon tumor or twisted bowel (volvulus) needs to be always considered and properly treated, mostly surgically. Because constipation in patients with PD is so common and driven by multiple factors, it requires a thorough initial evaluation as well as frequent assessments to ensure improvement, stability, or deterioration. Periodic evaluations, using radiopaque markers and abdominal radiographs, are recommended to monitor patients with chronic constipation, even those who report no issues. Based on such surveillance assessments and testing, preventive measures can be implemented to avoid progressive dilation, atony, and loss of the propulsive activity of the colon. Similar steps may need to be taken in patients with PD who report bowel urgency, seepage, or stool incontinence, since that may be a hidden manifestation of stagnation of stool in the rectum, rectal distention, and intermittent release of mucus and fluid through the anal canal at unpredictable times, simulating diarrhea (Fig. 3.9).

Diarrhea

Diarrhea is defined as the passage of three of more loose stools per day. Based on the duration of symptoms, diarrhea is further classified as acute (<14 days) or chronic (>30 days). The most common causes of acute diarrhea are bacterial, such as *Salmonella*, *Campylobacter*, Shiga toxin-producing *E. coli*, *Vibrio*, and *Yersinia*. *Clostridium difficile* infection is an important cause of diarrhea following antibiotic or immunosuppressant treatment and in hospitalized patients. Antibiotic-associated diarrhea occurs in the absence of *Clostridium difficile* infection and is probably due to small intestinal bacterial overgrowth (SIBO).

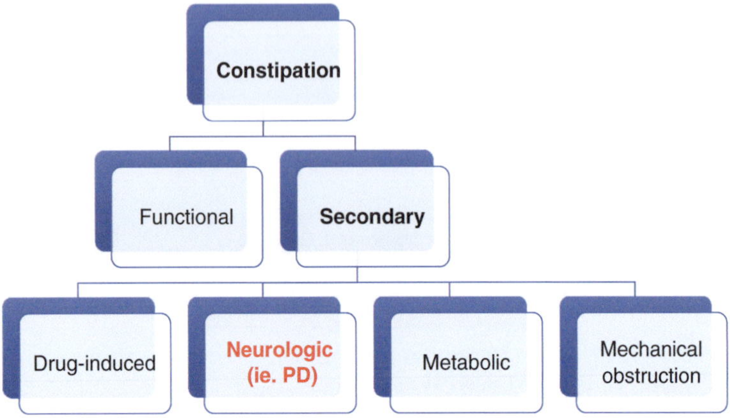

Fig. 3.9 The various etiologies for constipation, highlight a multifactorial pathophysiology in patients with PD that extends beyond disease-related neurodegeneration

When assessing a PD patient with chronic intermittent diarrhea, it is essential to clearly distinguish it from fecal incontinence associated with fecal impaction and overflow, a much more common issue. In turn, if true diarrhea is suspected, it is practical to classify it as "osmotic" (due to impaired fluid absorption, resulting from malabsorption, which improves with fasting) and "secretory" (excessive liquid stools >1 L/day, usually due to infection, which persists with fasting). Frequent, small-volume stools suggest a colonic origin, whereas less frequent but large-volume, often watery stools indicate a possible small bowel source.

Because PD is a disease of older adults, the presence of frequent, small, usually formed stools should suggest colonic source, such as colorectal cancer, diverticulosis, or solitary rectal ulcer, and a colonoscopy should always be considered. In contrast, frequent, watery bowel movements should raise the suspicion of small bowel bacterial overgrowth (SIBO), usually associated with pale, large-volume, foul-smelling stools with bloating, borborygmi and weight loss, or microscopic (lymphocytic or collagenous) colitis, particularly in women. In such instances, colonoscopy with multiple random biopsies should be considered.

In acute diarrhea, stool samples should be analyzed for parasites and bacteria. In chronic cases, stool should also be tested for elastase (low in pancreatic insufficiency) and calprotectin (elevated in any inflammatory condition and therefore useful to distinguish inflammatory from functional causes). Imaging studies (particularly MRI enterography) can provide additional useful information on small bowel anatomy and transit time which may be affected in PD.

There are several clues that assist the clinician in further clarifying the underlying diagnosis. Osmotic diarrhea improves with fasting, while secretory diarrhea, typically >1 L/day, persists with fasting. Frequent, small-volume stools suggest a large bowel origin (i.e., colorectal cancer, diverticular disease, solitary rectal ulcer), whereas less frequent, but large-volume, often watery stools indicate a possible small bowel origin (i.e., laxative, nonsteroidal anti-inflammatory drugs, or proton pump inhibitor use). Frequent, watery bowel movements without bleeding suggest collagenous or lymphocytic colitis, bile salt malabsorption, infection (i.e., giardiasis), lactose or sucrose intolerance, or diarrhea-predominant irritable bowel syndrome. Fatty stools that are difficult to flush away (steatorrhea) indicate pancreatic insufficiency. Pale, large-volume, foul-smelling stools with bloating and borborygmi and weight loss suggest small bowel bacteria overgrowth (SIBO). In contrast, the presence of blood suggests a colonic origin, such as colitis, colorectal cancer, or diverticular disease.

Bowel Urgency and Fecal Incontinence

Anorectal control (continence) is normally maintained by a combination of several factors: (1) the internal (involuntary) anal sphincter, (2) the external (voluntary) anal (striated) sphincter, (3) the puborectalis muscle, and (4) anorectal sensation. Weakness of internal anal (involuntary) sphincter leads to incontinence of flatus and soiling if the patient has diarrhea. Table outlines the value of manometric evaluation of patients with PD suffering from bowel urgency and fecal incontinence (Fig. 3.10).

A patient with PD who is affected by the inadvertent voiding of rectal contents through the anal canal (fecal incontinence) lives a life of social alienation and personal isolation. "Urge" incontinence is defined as the unwanted loss of stool despite active attempts to inhibit stool evacuation. "Passive" incontinence is defined as the unwanted loss of stool without the person being aware. Based on these definitions, there are three FI subtypes (urge, passive, mixed). Patients who report that they are almost exclusively incontinent because they have "urgency and cannot reach the toilet on time" are considered to have "urge" incontinence, whereas patients who are almost exclusively "unaware when the leakage is actually happening" are considered to have "passive" incontinence. Patients who report both are considered to have "mixed" incontinence [38]. It is essential to establish the degree of disability, since the management of the patient with partial soiling secondary to a prolapsed hemorrhoid will be very different from that of a patient with frequent incontinence of formed stool.

In the assessment of FI, its duration, details of stool consistency using the Bristol stool form scale [39], and the presence of transit disorders using the Rome IV criteria [40] should be collected using standard questionnaires at baseline and during

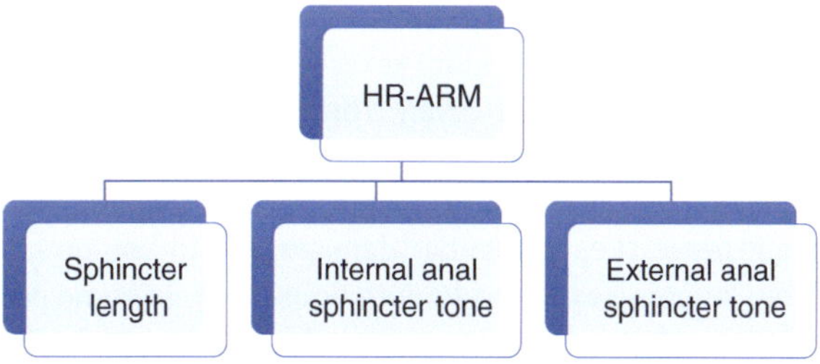

Fig. 3.10 Diagrammatic highlight of the essential components to be measured -and periodically reassessed over time- using high-resolution anorectal manometry (HR-ARM)

follow-up. FI severity is assessed using the Cleveland Clinic Incontinence Score (CCIS) [41] or the Vaizey score, which incorporates the use of constipating medications and the lack of ability to defer defecation for 15 min [42]. Quality of life (QOL) is usually measured using the American Society of Colon and Rectal Surgeons QOL auto-questionnaire for FI (FIQL) [43]. One of the items of the embarrassment domain of the FIQL questionnaire is related to passive FI ("I leak stool without even knowing it").

The external anal sphincter can vigorously contract upon volition for approximately 60 s to prevent soiling in case of the presence of loose stool in the rectum. Most of the anorectal control is provided by the contraction of the puborectalis muscle, which creates an angle between the lower rectum and upper anal canal (the anorectal angle). Sharp angulation creates a flap-valve mechanism to operate such that increases in intra-abdominal pressure cause the anterior rectal wall to close over the top of the anal canal, preventing rectal contents entering it. The sensation of a full rectum is probably caused by tension on pressure receptors in the pelvic floor musculature. The discrimination of the nature of rectal contents is accomplished by a locally mediated reflex whereby rectal distention (from flatus or stool) initiates internal anal sphincter relaxation. A sample of rectal contents therefore intrudes into the anal canal where it is perceived.

In all patients, a full clinical examination, perianal inspection, and dynamic digital rectal examination (at rest, upon squeeze and push maneuvers) should be enhanced by colonoscopy and high-resolution anorectal manometry to validate both the structural and functional integrity of the anorectum. In special cases, electromyography of the external anal sphincter and puborectalis muscles, anorectal ultrasound, and EndoFLIP measurements may be needed. MR defecography frequently provides valuable information on the integrity of the pelvic floor and the assessment of pelvic (uterine and bladder) and rectal prolapse. FI does not necessarily imply deficiency of the anal sphincters or the pelvic floor musculature, because variable degrees of soiling may occur normally, if a person experiences acute or chronic diarrhea. This is why it is important to structurally examine the rectum because patients with inflammatory bowel disease frequently state that the urgency and FI is the most distressing element of their disease. Older patients with PD may also have poor awareness of rectal filling resulting in fecal impaction. In these patients, the internal sphincter relaxes in response to rectal distention, facilitating leakage of loose stool.

Minor inadvertent loss of flatus or liquid stool through the anus typically results from a weak internal anal sphincter, and it may be associated with hemorrhoids or rectal prolapse. Major inadvertent and/or frequent loss of fully formed stool through the rectum suggests a more severe functional impairment of the anorectal structure and function and may result from worsening of preexisting surgical or obstetrical trauma of the external anal sphincter and the puborectalis muscle. Most of these patients have denervation of the striated (external) anal sphincter resulting from local nerve (i.e., pudendal) damage due to compression and stretching forces or from straining excessively with defecation. In some patients, partly because of sensory loss, fecal impaction ensues (Figs. 3.11 and 3.12).

Bowel (fecal) urgency	Sudden and immediate need to have a bowel movement
Fecal incontinence	Involuntary loss of solid or liquid stool
Fecal seepage	Leakage of liquid stool after a normal bowel movement
Gas-bowel uncertainty	Inability to distinguish need to pass stool or gas
Passive incontinence	Involuntary passing of stool without awareness
Stool frequency	Visits to the toilet to pass stool, mucus, or gas
Urge incontinence	Involuntary passing of stool due to inability to reach the toilet in time

Fig. 3.11 Clinical differentiation of various forms of fecal urgency and incontinence in patients with Gut PD to assist in further testing and management

Fig. 3.12 Impact of bowel urgency on quality of life and social function in PD

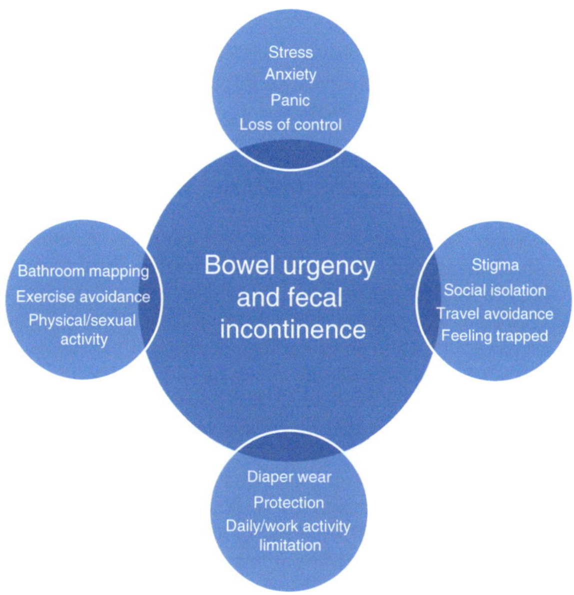

Rectal Bleeding

Typically, rectal bleeding in patients with PD occurs in the context of distal colonic pathology and associated constipation, but other, more ominous conditions, such as colorectal cancer, need to be carefully considered. Because of the long-standing gut symptoms, such as fullness and nausea associated with gastroparesis or constipation, patients may not be adequately prepared for a colonoscopy and malignancy may be missed. A good-quality colonoscopy is imperative in all cases and will provide an explanation as to the origin and significance of the bleeding. Typically, bleeding is related to anal fissure, stercoral ulceration, diverticulosis, and internal hemorrhoids. Unlikely etiologies include colonic ischemia and inflammatory bowel disease or infectious colitis. A recent provocative study has raised the possibility that PD may increase the risk of hematological pathology [44].

References

1. Yang CL, Huang JP, Tan YC, Wang TT, Zhang H, Qu Y. The effectiveness and safety of botulinum toxin injections for the treatment of sialorrhea with Parkinson's disease: a systematic review and meta-analysis. BMC Pharmacol Toxicol. 2023;24(1):52. https://doi.org/10.1186/s40360-023-00694-7. PMID: 37828600; PMCID: PMC10571401.
2. Verhoeff MC, Koutris M, Vries R, Berendse HW, Dijk KDV, Lobbezoo F. Salivation in Parkinson's disease: a scoping review. Gerodontology. 2023;40(1):26–38. https://doi.org/10.1111/ger.12628. Epub 2022 Mar 4.
3. Siau R, Kinshuck A, Houghton L. The assessment and management of globus pharyngeus. Br J Hosp Med (Lond). 2021;82(3):1–8. https://doi.org/10.12968/hmed.2021.0038. Epub 2021 Mar 31.
4. Van Daele DJ. Esophageal manometry, pH testing, endoscopy, and videofluoroscopy in patients with globus sensation. Laryngoscope. 2020;130(9):2120–5. https://doi.org/10.1002/lary.28289. Epub 2019 Sep 9.
5. Sung HY, Kim JS, Lee KS, Kim YI, Song IU, Chung SW, Yang DW, Cho YK, Park JM, Lee IS, Kim SW, Chung IS, Choi MG. The prevalence and patterns of pharyngoesophageal dysmotility in patients with early-stage Parkinson's disease. Mov Disord. 2010;25(14):2361–8. https://doi.org/10.1002/mds.23290.
6. Fasano A, Visanji NP, Liu LW, Lang AE, Pfeiffer RF. Gastrointestinal dysfunction in Parkinson's disease. Lancet Neurol. 2015;14(6):625–39. https://doi.org/10.1016/S1474-4422(15)00007-1.
7. Bushmann M, Dobmeyer SM, Leeker L, Perlmutter JS. Swallowing abnormalities and their response to treatment in Parkinson's disease. Neurology. 1989;39(10):1309–14. https://doi.org/10.1212/wnl.39.10.1309.
8. Potulska A, Friedman A, Królicki L, Spychala A. Swallowing disorders in Parkinson's disease. Parkinsonism Relat Disord. 2003;9(6):349–53. https://doi.org/10.1016/s1353-8020(03)00045-2.
9. Lin Z, Nicodème F, Lin CY, Mogni B, Friesen L, Kahrilas PJ, Pandolfino JE. Parameters for quantifying bolus retention with high-resolution impedance manometry. Neurogastroenterol Motil. 2014;26(7):929–36. https://doi.org/10.1111/nmo.12346. Epub 2014 Apr 22. PMID: 24750336; PMCID: PMC4120956.
10. Bulsiewicz WJ, Kahrilas PJ, Kwiatek MA, Ghosh SK, Meek A, Pandolfino JE. Esophageal pressure topography criteria indicative of incomplete bolus clearance: a study using high-resolution impedance manometry. Am J Gastroenterol. 2009;104(11):2721–8. https://doi.org/10.1038/ajg.2009.467. Epub 2009 Aug 18. PMID: 19690527; PMCID: PMC2886600.
11. Kahrilas PJ, Bredenoord AJ, Fox M, Gyawali CP, Roman S, Smout AJ, Pandolfino JE, International High Resolution Manometry Working Group. The Chicago Classification of esophageal motility disorders, v3.0. Neurogastroenterol Motil. 2015;27(2):160–74. https://doi.org/10.1111/nmo.12477. Epub 2014 Dec 3. PMID: 25469569; PMCID: PMC4308501.
12. Kerschner A, Hassan H, Kern M, Edeani F, Mei L, Sanvanson P, Shaker R, Yu E. Parkinson's disease is associated with low striated esophagus contractility potentially contributing to the development of dysphagia. Neurogastroenterol Motil. 2024;36(8):e14822. https://doi.org/10.1111/nmo.14822. Epub 2024 May 26.

13. Mehanna R, Jankovic J. Respiratory problems in neurologic movement disorders. Parkinsonism Relat Disord. 2010;16(10):628–38. https://doi.org/10.1016/j.parkreldis.2010.07.004. Epub 2010 Aug 2.
14. Claus I, Muhle P, Czechowski J, Ahring S, Labeit B, Suntrup-Krueger S, Wiendl H, Dziewas R, Warnecke T. Expiratory muscle strength training for therapy of pharyngeal dysphagia in Parkinson's disease. Mov Disord. 2021;36(8):1815–24. https://doi.org/10.1002/mds.28552. Epub 2021 Mar 2.
15. Sawada A, Zhang M, Ustaoglu A, Nikaki K, Lee C, Woodland P, Yazaki E, Takashima S, Ominami M, Tanaka F, Ciafardini C, Nachman F, Ditaranto A, Agotegaray J, Bilder C, Savarino E, Gyawali CP, Penagini R, Fujiwara Y, Sifrim D. Superficial oesophageal mucosal innervation may contribute to severity of symptoms in oesophageal motility disorders. Aliment Pharmacol Ther. 2024;59(1):100–12. https://doi.org/10.1111/apt.17773. Epub 2023 Oct 16.
16. Triadafilopoulos G, Mashimo H, Tatum R, O'Clarke J, Hawn M. Mixed esophageal disease (MED): a new concept. Dig Dis Sci. 2023;68(9):3542–54. https://doi.org/10.1007/s10620-023-08008-x. Epub 2023 Jul 20.
17. Simadibrata DM, Lesmana E, Fass R. Role of endoscopy in gastroesophageal reflux disease. Clin Endosc. 2023;56(6):681–92. https://doi.org/10.5946/ce.2023.182. Epub 2023 Oct 12. PMID: 37822063; PMCID: PMC10665616.
18. Triadafilopoulos G. Endoscopic non-ablative radiofrequency treatment (Stretta) for gastroesophageal reflux disease (GERD). Curr Gastroenterol Rep. 2024;26(4):93–8. https://doi.org/10.1007/s11894-024-00917-9. Epub 2024 Feb 23.
19. Salvador R, Vittori A, Capovilla G, Riccio F, Nezi G, Forattini F, Provenzano L, Nicoletti L, Moletta L, Costantini A, Valmasoni M, Costantini M. Antireflux surgery's lifespan: 20 years after laparoscopic fundoplication. J Gastrointest Surg. 2023;27(11):2325–35. https://doi.org/10.1007/s11605-023-05797-4. Epub 2023 Aug 14. PMID: 37580489; PMCID: PMC10661768.
20. Wyman JB, Dent J, Heddle R, et al. Control of belching by the lower esophageal sphincter. Gut. 1990;31:639–46.
21. Straathof JW, Ringers J, Lamers CB, et al. Provocation of transient lower esophageal sphincter relaxations by gastric distention with air. Am J Gastroenterol. 2001;8:2317–23.
22. Lin M, Triadafilopoulos G. Belching: dyspepsia or gastroesophageal reflux disease? Am J Gastroenterol. 2003;98(10):2139–45. https://doi.org/10.1111/j.1572-0241.2003.07627.x.
23. Wei BR, Zhao YJ, Cheng YF, Huang C, Zhang F. *Helicobacter pylori* infection and Parkinson's disease: etiology, pathogenesis and levodopa bioavailability. Immun Ageing. 2024;21(1):1. https://doi.org/10.1186/s12979-023-00404-1. PMID: 38166953; PMCID: PMC10759355.
24. Chang JJ, Gadi SRV, Videnovic A, Kuo B, Pasricha TS. Impact of outpatient gastroenterology consult on pharmacotherapy and management of gastrointestinal symptoms in Parkinson's disease. Clin Park Relat Disord. 2023;9:100215. https://doi.org/10.1016/j.prdoa.2023.100215. PMID: 37700817; PMCID: PMC10493246.
25. Wysokiński A, Sobów T, Kłoszewska I, Kostka T. Mechanisms of the anorexia of aging—a review. Age (Dordr). 2015;37(4):9821. https://doi.org/10.1007/s11357-015-9821-x. Epub 2015 Aug 1. PMID: 26232135; PMCID: PMC5005824.
26. Triadafilopoulos G, Lee M, Neshatian L. High prevalence of anorectal dysfunction in ambulatory patients with chronic constipation, regardless of colon transit time. Dig Dis Sci. 2024;69(1):180–8. https://doi.org/10.1007/s10620-023-08072-3. Epub 2023 Aug 9.
27. Tan AH, Mahadeva S, Thalha AM, Gibson PR, Kiew CK, Yeat CM, Ng SW, Ang SP, Chow SK, Tan CT, Yong HS, Marras C, Fox SH, Lim SY. Small intestinal bacterial overgrowth in Parkinson's disease. Parkinsonism Relat Disord. 2014;20(5):535–40. https://doi.org/10.1016/j.parkreldis.2014.02.019. Epub 2014 Mar 2.
28. Jensen GL, Cederholm T, Correia MITD, Gonzalez MC, Fukushima R, Higashiguchi T, de Baptista GA, Barazzoni R, Blaauw R, Coats AJS, Crivelli A, Evans DC, Gramlich L, Fuchs-Tarlovsky V, Keller H, Llido L, Malone A, Mogensen KM, Morley JE, Muscaritoli M, Nyulasi

I, Pirlich M, Pisprasert V, de van der Schueren M, Siltharm S, Singer P, Tappenden KA, Velasco N, Waitzberg DL, Yamwong P, Yu J, Compher C, Van Gossum A. GLIM criteria for the diagnosis of malnutrition: a consensus report from the Global Clinical Nutrition Community. JPEN J Parenter Enteral Nutr. 2019;43(1):32–40. https://doi.org/10.1002/jpen.1440. Epub 2018 Sep 2.

29. Zawada ET Jr. Malnutrition in the elderly. Is it simply a matter of not eating enough? Postgrad Med. 1996;100(1):207–8. https://doi.org/10.3810/pgm.1996.07.17.
30. Edwards LL, Quigley EM, Pfeiffer RF. Gastrointestinal dysfunction in Parkinson's disease: frequency and pathophysiology. Neurology. 1992;42(4):726–32. https://doi.org/10.1212/wnl.42.4.726.
31. Evans WJ, Morley JE, Argilés J, Bales C, Baracos V, Guttridge D, Jatoi A, Kalantar-Zadeh K, Lochs H, Mantovani G, Marks D, Mitch WE, Muscaritoli M, Najand A, Ponikowski P, Rossi Fanelli F, Schambelan M, Schols A, Schuster M, Thomas D, Wolfe R, Anker SD. Cachexia: a new definition. Clin Nutr. 2008;27(6):793–9. https://doi.org/10.1016/j.clnu.2008.06.013. Epub 2008 Aug 21.
32. Fried LP, Cohen AA, Xue QL, Walston J, Bandeen-Roche K, Varadhan R. The physical frailty syndrome as a transition from homeostatic symphony to cacophony. Nat Aging. 2021;1(1):36–46.
33. Stenholm S, Ferrucci L, Vahtera J, Hoogendijk EO, Huisman M, Pentti J, Lindbohm JV, Bandinelli S, Guralnik JM, Kivimäki M. Natural course of frailty components in people who develop frailty syndrome: evidence from two cohort studies. J Gerontol A Biol Sci Med Sci. 2019;74(5):667–74.
34. Perazza LR, Brown-Borg HM, Thompson LV. Physiological systems in promoting frailty. Compr Physiol. 2022;12(3):3575–620.
35. Ferrucci L, Fabbri E. Inflammageing: chronic inflammation in ageing, cardiovascular disease, and frailty. Nat Rev Cardiol. 2018;15(9):505–22.
36. Morley JE, Malmstrom TK, Miller DK. A simple frailty questionnaire (FRAIL) predicts outcomes in middle aged African Americans. J Nutr Health Aging. 2012;16(7):601–8.
37. Segers K, Surquin M. Can mirtazapine counteract the weight loss associated with Alzheimer disease? A retrospective open-label study. Alzheimer Dis Assoc Disord. 2014;28(3):291–3. https://doi.org/10.1097/WAD.0b013e3182614f52.
38. Abrams P, Andersson KE, Birder L, Brubaker L, Cardozo L, Chapple C, et al. Fourth international consultation on incontinence recommendations of the international scientific committee: evaluation and treatment of urinary incontinence, pelvic organ prolapse, and fecal incontinence. Neurourol Urodyn. 2010;29:213–40. https://doi.org/10.1002/nau.20870.
39. Lewis SJ, Heaton KW. Stool form scale as a useful guide to intestinal transit time. Scand J Gastroenterol. 1997;32:920–4. https://doi.org/10.3109/00365529709011203.
40. Mearin F, Lacy BE, Chang L, Chey WD, Lembo AJ, Simren M, et al. Bowel disorders. Gastroenterology. 2016;150:1393–407. https://doi.org/10.1053/j.gastro.2016.02.031.
41. Jorge JM, Wexner SD. Etiology and management of fecal incontinence. Dis Colon Rectum. 1993;36:77–97. https://doi.org/10.1007/BF02050307.
42. Vaizey CJ, Carapeti E, Cahill JA, Kamm MA. Prospective comparison of faecal incontinence grading systems. Gut. 1999;44:77–80.
43. Rockwood TH, Church JM, Fleshman JW, Kane RL, Mavrantonis C, Thorson AG, Wexner SD, Bliss D, Lowry AC. Fecal incontinence quality of life scale: quality of life instrument for patients with fecal incontinence. Dis Colon Rectum. 2000;43(1):9–16. https://doi.org/10.1007/BF02237236. PMID: 10813117.
44. Shahrestani S, Gendreau J, Tafreshi AR, Brown NJ, Dashtipour K. High rates of blood transfusion associated with Parkinson's disease. Neurol Sci. 2022;43(8):4761–8. https://doi.org/10.1007/s10072-022-06097-6. Epub 2022 May 2. PMID: 35499631; PMCID: PMC9349070.

4 Diagnostic and Assessment Tools

General

Gut Symptoms and Study Tools in Gut PD

Patients with gut PD may exhibit either an isolated symptom of gastrointestinal dysfunction that brought them to clinical attention or a multitude of symptoms, suggestive of multifocal gut involvement. Managing the presenting symptom alone may be inadequate and unsuccessful or may complicate other symptoms stemming from other areas of gut involvement [1]. Depending on the clinical presentation, the various ancillary tests outlined below may be used in isolation or often in conjunction with each other, since they may provide complementary information that will guide therapy. Since the disease is often progressive, periodic, multimodality evaluations and reassessments are recommended [2]. These study tools are variably available, and they require guidance and counseling by a gastroenterologist specializing in neurogastroenterology, typically in referral centers. Test justification, ordering, and interpretation of the results will be in turn coordinated and discussed with the patient's movement disorder neurologist, and a concerted effort will be initiated to maximize any gains from therapy (Fig. 4.1).

Tools

The tools highlighted below may be variably used by physician specialists, as indicated. Each result is shared and ultimately placed in the context of the overall neurological picture, symptom severity, and quality of life impact. The movement disorder specialist remains at the helm of coordination efforts, exchanging with other health team members for a synthesis of the findings and to coordinate action plan.

G. Triadafilopoulos, *The Gut in Parkinson's Disease*,
https://doi.org/10.1007/978-3-031-77171-2_4

Symptom	Tool of study
Dental problems	Dental examination
Drooling	Sialometry; video-fluoroscopy
Dry mouth	Medication use history; sialometry; salivary scintigraphy
Lump in throat (globus)	Flexible endoscopic evaluation of swallowing (FEES); Video fluoroscopic swallowing study (VFSS); barium esophagram
Dysphagia (difficulty swallowing)	Flexible endoscopic evaluation of swallowing (FEES); Video fluoroscopic swallowing study (VFSS); timed barium and tablet esophagram; endoscopy; high -resolution esophageal manometry; EnfoFLIP
Odynophagia (painful swallowing)	Endoscopy
Chest pain	High-resolution esophageal manometry (HREM)
Regurgitation	Endoscopy; barium esophagram; high -resolution esophageal manometry; EnfoFLIP; ambulatory pH monitoring
Heartburn	Endoscopy; ambulatory esophageal pH monitoring
Loss of appetite	Interview
Nausea and vomiting	Endoscopy; Stomach (gastric) emptying scan; ^{13}C-octanoate breath test; Ultrasonography; Electrogastrography; Wireless motility capsule; MR enterography; CT of abdomen with oral contrast
Early satiety (fullness after meals)	Endoscopy; stomach (gastric) emptying scan; Wireless motility capsule, Ultrasonography; MR enterography; CT of abdomen with oral contrast
Bloating and abdominal distention	Lactulose breath test; wireless motility capsule; MR enterography; CT of abdomen with oral contrast
Frequent burping (belching)	Endoscopy; Stomach (gastric) emptying scan
Constipation	Colonoscopy; wireless motility capsule; radio -opaque marker (Sitzmarks) test; high -resolution anorectal manometry; MR defecography
Abdominal pain	CT scan of abdomen & pelvis; plain abdominal radiograph; MR enterography
Bowel urgency and fecal (stool) incontinence	Rectal examination; Colonoscopy; high -resolution anorectal manometry; MR defecography

Fig. 4.1 Diagnostic tools for the evaluation of gastrointestinal symptoms in patients with Gut PD

Clinical

GI symptom questionnaires in gut PD

Structural Assessment

Endoscopic

Flexible endoscopic evaluation of swallowing (FEES)

Upper endoscopy

Video capsule endoscopy

Colonoscopy

Imaging

Salivary scintigraphy

Video-fluorography or modified barium swallow study

Timed barium (TBE) and barium tablet esophagram (BTE)

Plain abdominal X-ray (AXR)

Radiopaque marker study (Sitzmarks)

CT of abdomen and pelvis (CTAP)

MRI abdomen and pelvis

MR enterography

MR defecography

Abdominal ultrasound

Gastric Alimetry

Gastric emptying test (scintigraphy)

Functional Assessment

High-resolution esophageal manometry (HREM)

Antroduodenal manometry (ADM)

High-resolution anorectal manometry (HR-ARM), rectal sensation test (RST), and balloon expulsion test (BET)

Endoluminal functional lumen imaging probe (EndoFLIP)

24-h impedance-pH monitoring

96-h wireless esophageal pH monitoring

Laryngo-esophageal pH monitoring (Dx–pH measuring system)

Wireless motility capsule (WMC; SmartPill®)

Electrogastrography (EGG)

Others

^{13}C-octanoate breath test

Glucose and lactulose breath tests

Sialometry

Autonomic function tests

Clinical

Gastrointestinal Symptom Questionnaire in Gut PD

The use of a gastrointestinal symptom questionnaire is highly recommended before every visit to the physician, regardless of their specialty. The patient has the time to examine if they have any symptoms and to what degree such symptoms affect their quality of life by grading their frequency and severity. Relaying this information to the physician will then initiate a conversation and decision-making. The physician will be alerted to the problem and will decide about its management, specifically by an intervention, or a referral. The questionnaire can be used also by the patient, week after week, month after month, to assess how they are doing overall, with or without any intervention. Changes in the nature, frequency, and severity of symptoms will in turn allow the recognition of a new problem or the improvement or worsening of an existing problem. Importantly, the use of such questionnaire focuses the discussion and sets goals to be met by a particular intervention. The questionnaire also facilitates the communication between the primary physician, the neurologist, and the gastroenterologist so they can coordinate care, adjust medical therapy, or ensure ongoing monitoring.

The questionnaire has been particularly easy to fill out by the patient and their caretakers and provides very useful information to their physicians in a time-efficient manner. At the time of the visit, and with a brief glance, the physician can quickly decide how significant the gastrointestinal symptoms are, if they are focal or general in nature, and to what degree they are linked to the patient's neurological presentation, or potentially reflect side effects of therapy. If an intervention is decided upon, the questionnaire can be used as an assessment tool in judging its success or failure, the need for further testing, or advancement of therapy [3].

Gastrointestinal Symptom Questionnaire

Please **circle** your response to the questions below. Use the following **scale** to score your symptoms (Fig. 4.2):

0 No problem
1 Problem of **mild** intensity (<3/10) and/or frequency (<50% of the time)
2 Problem of **moderate** intensity (3–6/10) and/or frequency (50–75% of the time)
3 Problem of **severe** intensity (>7/10) and/or frequency (>75% of the time)

Symptom	Score			
Difficulty swallowing (dysphagia)	0	1	2	3
Chest pain with swallowing (odynophagia)	0	1	2	3
Fluid (acid or bile) regurgitation	0	1	2	3
Pain in upper stomach (above belly button)	0	1	2	3
Is this pain relieved with food or acid blockers	0	1	2	3
Is this pain worse with greasy foods	0	1	2	3
Is this pain waking you up at night	0	1	2	3
Pain in lower stomach (below belly button)	0	1	2	3
Is this pain waking you up at night?	0	1	2	3
Is this pain better with passing stool or gas?	0	1	2	3
Burping (belching)	0	1	2	3
Loss of appetite	0	1	2	3
Nausea (stomach sickness)	0	1	2	3
Early satiety (fullness after meals)	0	1	2	3
Bloating	0	1	2	3
Abdominal distention	0	1	2	3
Constipation (hard and infrequent) stool	0	1	2	3
Straining at stool evacuation	0	1	2	3
Unable to complete bowel movement	0	1	2	3
Urgency to pass stool	0	1	2	3
Stool (fecal) incontinence, or loss of bowel control	0	1	2	3
Diarrhea (loose and frequent) stool	0	1	2	3
Bleeding through the rectum	0	1	2	3
Unintentional weight loss (>10 lbs)	0	1	2	3

Fig. 4.2 Practical gastrointestinal symptom questionnaire to be used in patients with Gut PD. The questionnaire, filled out by the patient or caretaker, is then reviewed for clarity by the gastroenterologist and compared to any previous ones. It is a useful tool to monitor disease evolution and response to therapy

Structural Assessment

Flexible Endoscopic Evaluation of Swallowing (FEES)

A fiber-optic, full color endoscopic evaluation of swallowing (FEES) test is a procedure used to assess how well the patient swallows, and it requires a cooperative individual. During the procedure, a speech-language pathologist (SLP) passes a thin, flexible instrument through the nose under topical anesthesia and examines exclusively the nasopharynx, pharynx, and larynx, using normal liquids and foods in contrast to a barium examination. The examiner may use water, thicker liquids, pureed foods, soft foods, and even regular foods, noting for any problems chewing, swallowing, or breathing. They will also check if the voice sounds "wet," typically a sign of aspiration. FEES is a key method in swallowing evaluation, but its interpretation and the deduction of its clinical implications may be challenging. A recent review discussed an integrated FEES report that systematically addresses the endoscopic findings in relation to dysphagia severity, the various types of swallowing impairments, and the underlying mechanisms involved. In this process, validated scales and scores are used, findings are given clinical context, differential diagnoses are considered, further diagnostic steps are proposed, treatment options are evaluated, and a timeframe for reassessment is suggested. This framework is designed to be adaptable and open to continuous evolution [4].

Endoscopy

Endoscopy should be considered as first step in the evaluation of patients with gut PD who present with dysphagia, regurgitation, early satiety, and nausea, among many other symptoms suggestive of esophageal or gastric dysfunction, mostly to exclude structural abnormalities, such as esophagitis, *H. pylori* gastritis, peptic ulcer disease, or cancer (Fig. 4.3).

Below are some examples of esophageal abnormalities at upper endoscopy, typically seen in patients with PD. Left panel: White-yellow patches coating the surface of the esophagus that are characteristic of *Candida* esophagitis (a fungal infection). Identification of this infection, which typically causes dysphagia and painful swallowing, allows therapy with fluconazole 200 mg loading dose, followed by 100–200 mg daily for 7–14 days. In patients with PD, *Candida* esophagitis results from underlying ineffective esophageal motility (IEM) or distal esophageal spasm, causing stagnation (stasis) and poor esophageal clearance. Middle panel: Food bolus impaction causing acute inability to swallow (dysphagia) requiring expedited endoscopic removal. This is a medical emergency to avoid risk of complications, particularly aspiration pneumonia, because such patients are unable to swallow saliva. Right panel: Distal esophageal spasm recognized during an endoscopy. Medical therapy using an antispasmodic hyoscyamine 0.25 mg, under the tongue, three times daily, as needed, can be used. For more severe cases, endoscopically applied botulinum toxin injection typically provides short-term symptom relief (i.e., approximately 6 months), with low risk of adverse effects (Fig. 4.4).

Representative upper endoscopy images of the stomach and duodenum seen in patients with PD. Left panel: *H. pylori*-related ulcer (white crater, surrounded by inflammation of the duodenal bulb). This ulcer will require therapy against *H. pylori* infection and is expected to heal within 2–3 months. Middle panel: Fluid residual in the stomach lumen (green semiliquid material), obliterating a full visualization of

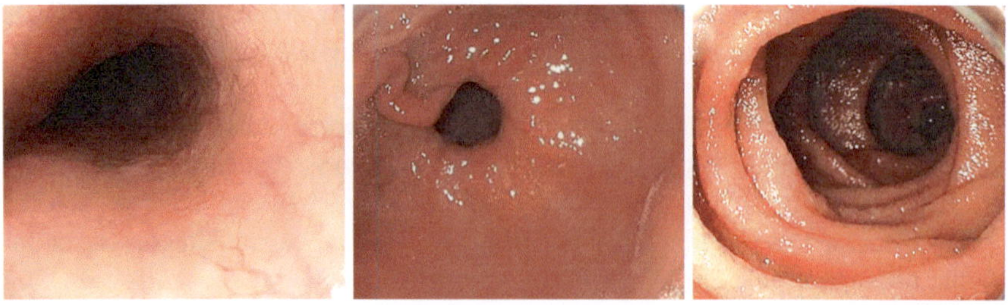

Fig. 4.3 The images exemplify the normal appearance of the esophagus (left), the stomach antrum and pylorus (middle), and duodenum (right)

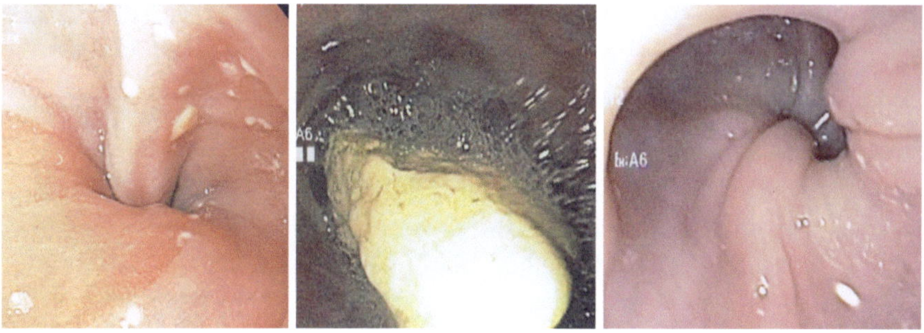

Fig. 4.4 White-yellow patches coating the surface of the esophagus that are characteristic of Candida esophagitis (left). Food bolus impaction (middle). Distal esophageal spasm recognized during an endoscopy (right)

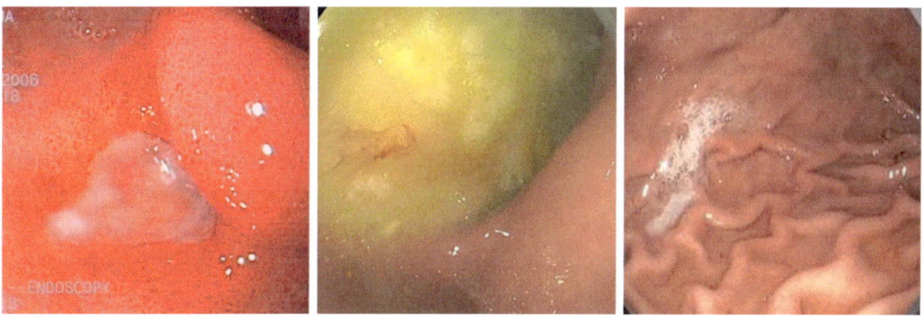

Fig. 4.5 Left panel: *H. pylori*-related ulcer (white crater, surrounded by inflammation of the duodenal bulb). Middle panel: Fluid residual in the stomach lumen, suggestive of gastroparesis. Right panel: Normal appearance of the stomach lining after prolonged fasting and aspiration of residual stomach contents on the same patient as in middle panel

the underlying lining, suggestive of gastroparesis. Prokinetic therapy or Botox injection of the pylorus alleviates symptoms and eventually allows better visualization of the entire stomach lining. Right panel: Normal appearance of the stomach lining after prolonged fasting and extensive aspiration of residual stomach contents on the same patient as in middle panel (Fig. 4.5).

Upper endoscopy images of the stomach seen in a 55-year-old woman with PD and intermittent dyspepsia and fullness after meals. Left panel: Food residual (bezoar) despite 12-h fasting in the stomach lumen (green semisolid material), obliterating a full visualization of the underlying lining, suggestive of gastroparesis. Visualization of the entire lumen of the stomach is impaired, and a repeat endoscopy after prolonged fasting or liquid diet for several days is needed. This is a nonspecific finding since it may occur in patients with diabetes mellitus or in patients using narcotics. Regardless, it predisposes patients to nighttime aspiration and may contribute to ongoing malnutrition and cachexia due to early satiety and postprandial fullness. Right panel: Normal appearance of the stomach lining after prolonged fasting and extensive aspiration of residual stomach contents (Fig. 4.6).

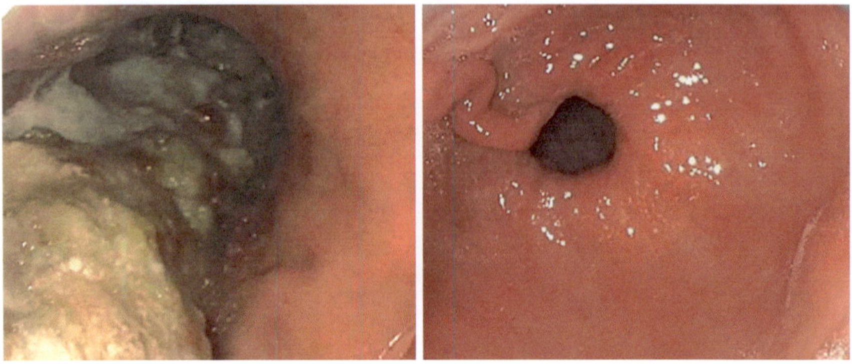

Fig. 4.6 Upper endoscopy images of the stomach. Left panel: Food residual (bezoar), suggestive of gastroparesis. Right panel: Normal appearance of the stomach of the same patient after prolonged fasting and extensive aspiration of residual stomach contents. The antrum and pylorus are well-seen

Video Capsule Endoscopy (VCE)

The video capsule, which is not larger than a vitamin pill, contains a small camera that takes thousands of images of the lumen of the digestive tract over 8 h or until the capsule is passed in the toilet after a bowel movement, whichever comes first. VCE is mainly used to view the small intestine which is difficult to access by endoscopy or colonoscopy. The images taken by the camera as it moves downstream from the small bowel contractions are transmitted to a special belt and stored by a recorder. When the study is completed, the pictures are downloaded to a computer, and software turns them into a video that is then reviewed by a physician. Capsule endoscopy is a safe and painless procedure, but it carries a slight risk for small bowel obstruction if the bowel has narrowed areas (strictures). Representative images of normal small bowel appearance on VCE are shown below (Fig. 4.7).

Colonoscopy

A colonoscopy should be considered as first step in the evaluation of patients with gut PD who present with constipation, bloating abdominal pain, and changes in bowel habits, among many other symptoms suggestive of large bowel dysfunction, to exclude structural abnormalities, such as colitis, stercoral ulcerations, diverticulosis, or cancer.

The images below show normal colonic mucosal lining. However, there is thickened sigmoid colon (myochosis), associated with chronic constipation. This is a nonspecific finding at colonoscopy, and it may occur in isolation or with associated diverticulosis. It contributes to left lower quadrant pain that relates to bowel

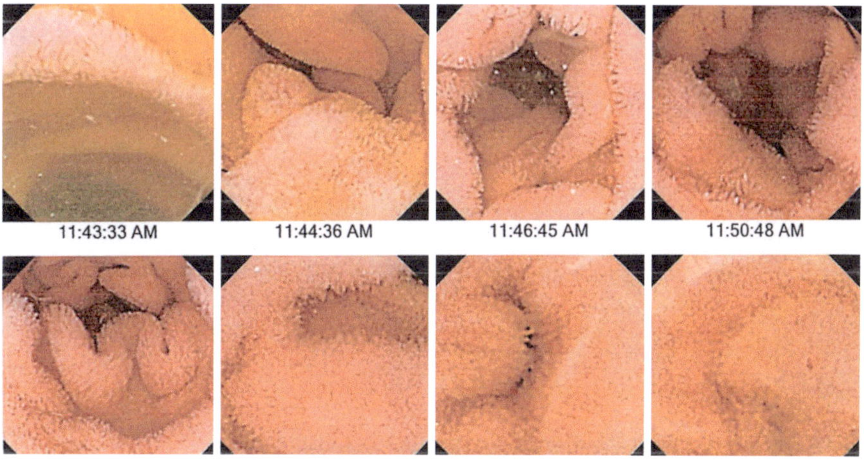

Fig. 4.7 Normal small bowel appearance on Video Capsule Endoscopy

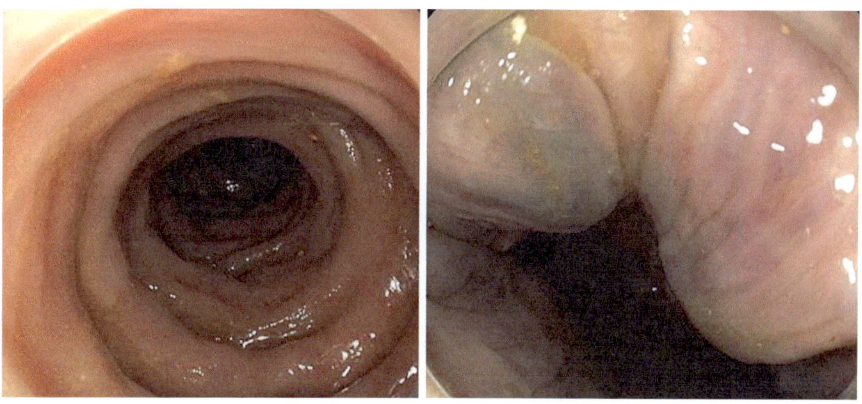

Fig. 4.8 Left panel: Normal colonic mucosal lining on colonoscopy. Right panel: Internal hemorrhoids

evacuation which typically occurs with significant difficulty and straining. Passing the colonoscope through this functional stenosis may be challenging (Fig. 4.8).

Figure (right) shows internal hemorrhoids noted at colonoscopy in a patient with PD who presented with chronic constipation, defecatory straining, and episodes of rectal bleeding. Most patients with PD are at an age group that requires screening colonoscopy to rule out colon polyps or cancer and appropriate surveillance as needed. In patients with rectal bleeding, one should always ascertain the absence of a structural abnormality. Although colorectal cancer is the most ominous diagnosis, other benign entities, such as stercoral ulcers, rectal prolapse, anal fissures, and internal hemorrhoids (above) can be seen at colonoscopy and properly managed (Figs. 4.9, 4.10, 4.11, 4.12).

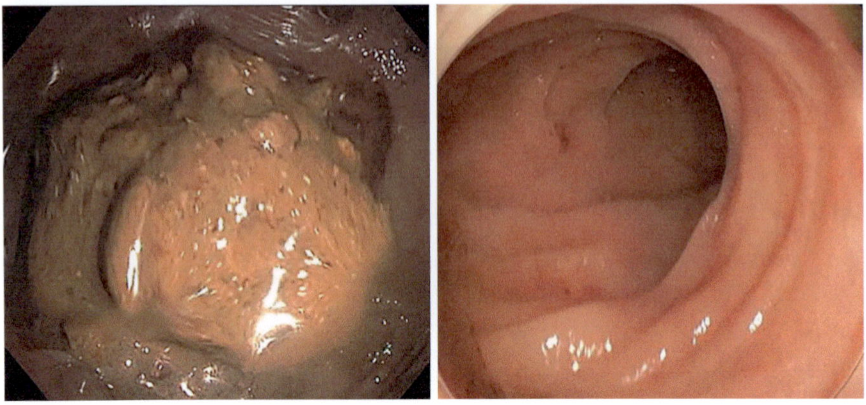

Fig. 4.9 Colonoscopy images of the rectum seen in an 82-year-old man with PD who was hospitalized because of painless rectal bleeding and anemia. Left panel: Stool residual impacted in the rectum lumen (brown material), obliterating a full visualization of the underlying lining, suggestive of fecal impaction. This was noted despite a full colonic lavage before the colonoscopy. Right panel: Normal appearance of the rectum lining after prolonged (2 consecutive days) colonic lavage using oral osmotic laxatives, administration of local enemas, and extensive aspiration of residual liquid colonic contents during the procedure

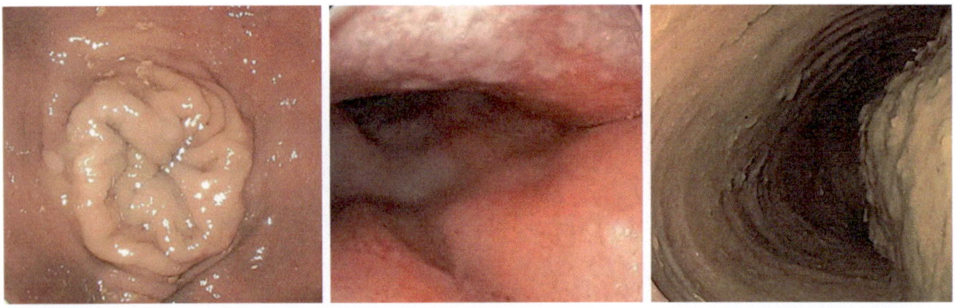

Fig. 4.10 Left panel depicts a colonoscopic view of the proximal rectum revealing internal rectal prolapse compromising transit through the rectum toward the anal canal in a 63-year-old white female with PD tremors, who presented with severe constipation and evacuation disorder. Middle panel shows a colonoscopic view of the rectum of a 76-year-old man with PD, revealing mucosal ischemic changes of rectal prolapse. Right panel shows a colonoscopic view of the rectum of an 88-year-old man with chronic, slow transit constipation and evacuation disorder. The visualization of the rectum is poor due to the presence of a fecal impaction due to inspissated stool, but there is no gross evidence of mucosal damage

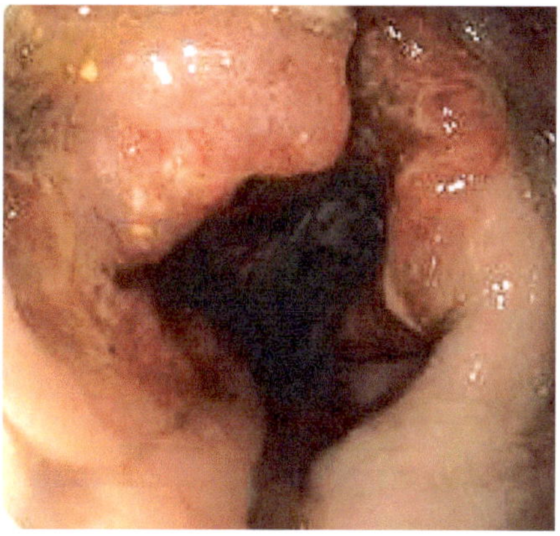

Fig. 4.11 This man was diagnosed with constipation 15 years ago, at age 62. A screening colonoscopy at that time was normal; he was advised to return for screening colonoscopy at age 72. PD was diagnosed 8 years ago and 7 years after the onset of constipation, at age 69, based on progressive rigidity and tremor. Focusing on his PD and because of impaired mobility, he did not follow up with the recommendation for repeat colonoscopy screening, maintaining his constipation under good control with lubiprostone 24 mcg orally twice daily. He presented now, at age 77, because of altered bowel habits and urge fecal incontinence. A colonoscopy showed a rectosigmoid cancer partially obstructing the lumen and contributing to his worsening symptom complex. In addition, sigmoid diverticulosis was noted proximal to the tumor, likely a result of long-standing constipation

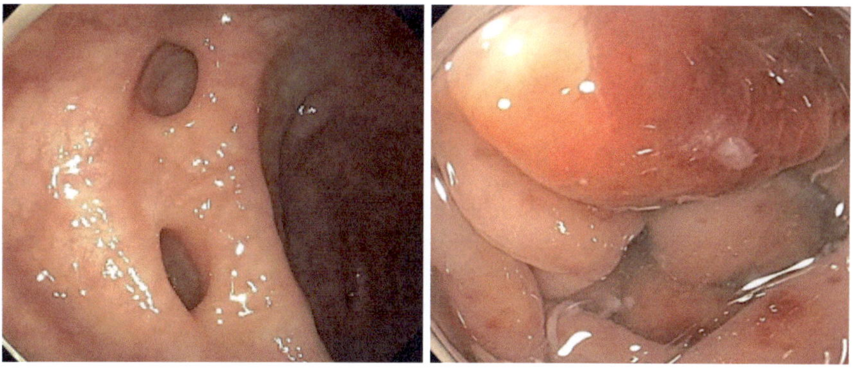

Fig. 4.12 An 83-year-old man with PD, bloating, left lower abdominal pain relieved with bowel evacuation, and long-standing constipation was found on colonoscopy to have sigmoid colon diverticulosis (left panel). Further advancement of the colonoscope identified an area of diverticula-associated colitis (DAC) (right panel) limiting access to the more proximal colon due to the underlying inflammation. Therapy with oral mesalamine was introduced with subsequent clinical improvement of the inflammation. Long-term osmotic laxatives were also used to effectively treat the patient's constipation

Imaging

Salivary Scintigraphy

Salivary gland scintigraphy is a nuclear medicine study that allows for functional evaluation of the salivary glands by assessing the uptake and excretion characteristics of radiotracer (intravenously injected technetium-99m pertechnetate) by salivary glands following secretive stimulation. It is not commonly used unless patients have severe xerostomia.

Video-Fluorography or Modified Barium Swallow Study

The video-fluorographic swallowing study (VFSS), also known as a modified barium swallowing examination (MBSS), is the test of choice for the real-time visualization of bolus flow *through* the upper aerodigestive tract that is both the breathing and swallowing systems. This test is particularly important in patients with PD who may suffer from either oropharyngeal, or esophageal dysphagia, or both, and it is frequently done to compliment the timed barium esophagram (TBE) and barium tablet esophagram (BTE). VFSS also permits detection of the presence and timing of aspiration, i.e., entry of ingested material through the level of the true vocal folds into the trachea, and identifies its causes by assessing the effects of various bolus volumes, bolus textures, and compensatory strategies on swallowing function. It identifies and distinguishes the type and severity of swallowing impairment and assesses the safety of oral intake, testing the impact of various interventions and formulating oral intake recommendations and treatments, such as gastric or enteral feedings (Fig. 4.13).

This patient developed slowly progressive, intermittent dysphagia and choking sensation after eating and particularly with pills. An upper endoscopy was normal as was a barium esophagram that failed to reveal a stricture. Upon video-fluorographic swallowing study, there was no aspiration into the larynx, and the oropharyngeal muscles were intact with liquid swallows that cleared the contrast through the esophagus. However, there was evidence of stagnation of the barium tablet in the valleculae that required at least ten swallowing efforts to clear.

Timed Barium Esophagram (TBE) and Barium Tablet Esophagram (BTE)

Conventional barium esophagram has long been considered the primary imaging modality of the esophagus. In the same fashion, a timed barium esophagram (TBE) is a valuable tool in the evaluation of esophagogastric junction outlet obstruction [5]. Timed barium esophagram is performed using an X-ray machine, with patient upright drinking 8 oz. or 236 mL of liquid barium (contrast agent), and the barium column height is evaluated at 1 and 5 min after ingestion. The barium column height

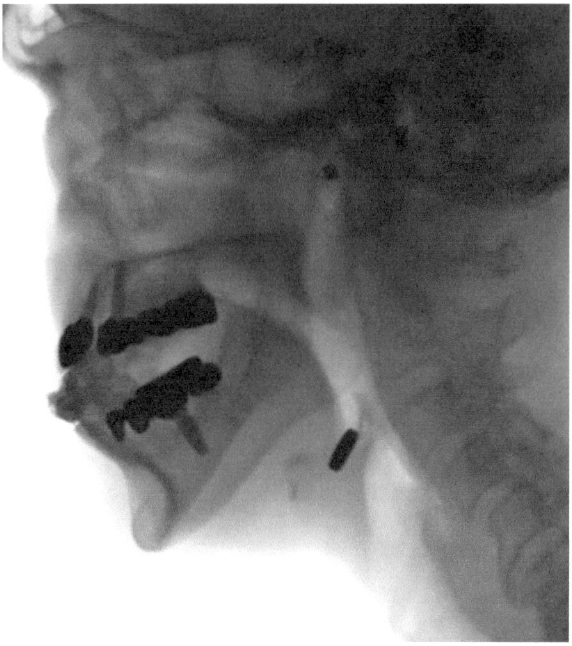

Fig. 4.13 Radiography of the neck shows barium tablet stagnating in the hypopharynx

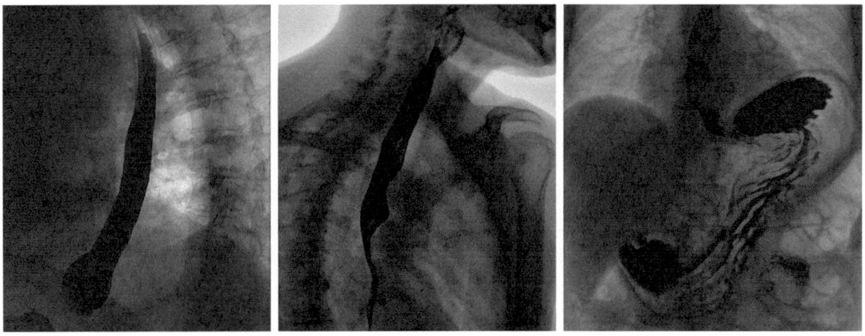

Fig. 4.14 Barium esophagram suggestive of esophageal body hypoperistalsis (ineffective esophageal motility, or IEM). Multiple views need to be reviewed to establish the diagnosis

supports esophageal outflow obstruction at the level where the esophagus meets the stomach, if >5 cm at 1 min or >2 cm at 5 min. Further, an improvement (decrease) in pre- to posttreatment (typically endoscopic Botox injection) barium height at 5 min might predict long-term clinical remission. Corkscrew appearance may be suggestive of distal esophageal spasm. Similarly, barium tablet esophagram is performed using an X-ray machine, with patient upright swallowing a 13-mm-diameter barium tablet which eventually dissolves and passes into the stomach. The 13-mm barium tablet retention again supports an obstructive process and poor esophageal clearance [6] (Fig. 4.14).

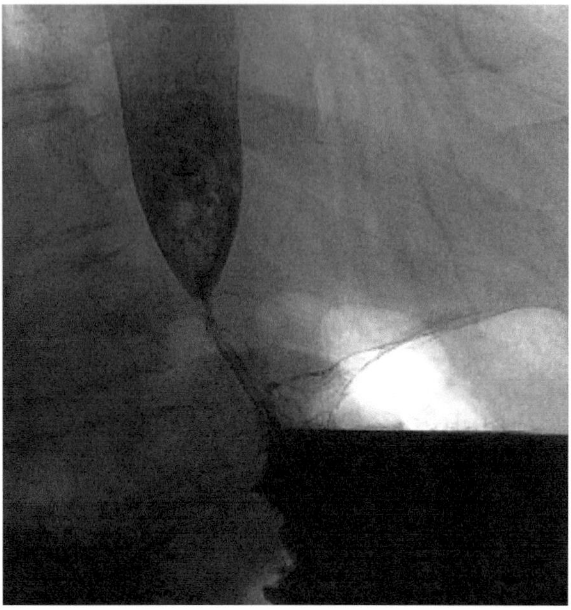

Fig. 4.15 Radiography of the gastroesophageal junction (EGJ) shows disintegrating barium tablet just above the junction, due to EGJ outlet obstruction (EGJOO), simulating esophageal achalasia

In this patient with PD and nonacid regurgitation, the swallowing initiation was normal; there was no aspiration. There was free flow of barium contrast into the esophagus (black cylindrical structure) without any mass or stricture defect. A barium tablet was easily passed into the stomach (not shown). Small hiatal hernia (distal ballooning of the esophagus) is present. The esophageal motility was normal. Contrast was emptied from the esophagus immediately and before the 1 min (Fig. 4.15).

In this patient with PD who presented with chronic heartburn and frequent dysphagia to solids and pills, a barium tablet esophagram revealed transient stasis of the barium tablet administered by mouth (whitish bubbles in distal esophagus represent the dissolving barium tablet) proximal to a distal stenosis (stricture). Endoscopic dilation and acid blockers were effective in resolving the symptoms.

Plain Abdominal X-Ray (AXR)

A plain abdominal X-ray (AXR) uses a very small dose of radiation to produce images of the abdominal cavity. It is used to evaluate the stomach, liver, the small intestine, the colon, and spleen to help diagnose unexplained abdominal pain, nausea, vomiting, diarrhea, or constipation. Because it is fast, inexpensive, and easy and requires no preparation, AXR is particularly useful in emergency diagnosis and treatment.

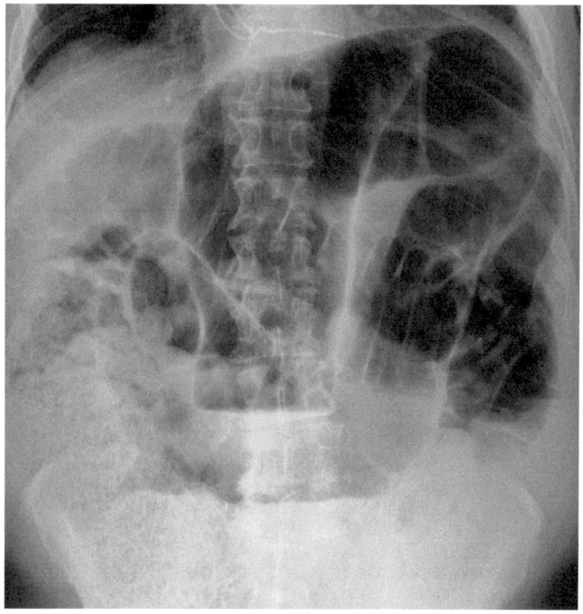

Fig. 4.16 Abdominal radiography showing abdominal distention due to stool (white areas in lower part of image) and gas (dark areas in upper part of image), mostly in the colon

In the example below, there is evidence of abdominal distention due to stool (white areas in lower part of image) and gas (dark areas in upper part of image), mostly in the colon, raising the possibility of obstruction, in this case related to fecal impaction from stool in the rectum because of poor tone of the colon and rectum in this patient with PD (Fig. 4.16).

In this AXR example, this patient with PD and severe constipation has had a prior anterior lumbar fusion with placement of a spacer device and anterior fixation hardware that are not displaced. Bone fusion is present across the L2–L3 intervertebral disk space, while the intervertebral disk space at L4–L5 is essentially obliterated, and there is a reduction in height of L5–S1. Lateral osteophyte formation is present on the right at L2–L3, extending laterally on either side of midline at L3–L4. Such disk osteophyte complex encroaches the L2 nerve root and encroaches on the exiting L3 roots on the left, while at L4–L5 it encroaches on the L4 root lateral to the foramen (not shown). Diffuse facet arthropathy is present in the lower lumbar region. There is no significant spinal canal or neural foraminal stenosis. These changes were causing significant back pain, requiring narcotics for relief and, in turn, contributing to the patient's chronic constipation, due to opioid use (opioid-induced constipation). The patient benefited in part by daily administration of naloxegol 25 mg, in addition to other combination therapies (Fig. 4.17).

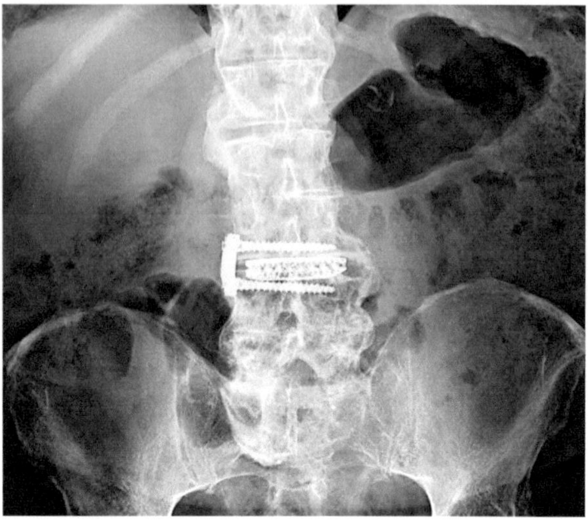

Fig. 4.17 Abdominal X-ray (AXR) showing evidence of increased colonic stool burden suggestive of constipation

Radiopaque Marker Study (Sitzmarks)

A Sitzmarks study is used with patients who are suffering from chronic constipation (typically less than two bowel movements per week). For the test, tiny "markers" are used to see how fast food is moving through the intestines. The patient swallows one gelatin capsule that contains tiny plastic ring markers which will show up on X-rays and lets the doctors track their progress through the intestines. After capsule ingestion, the patient is asked to make an appointment for an abdominal X-ray 5 days later to see if any of the "markers" are left in the colon or they have been expelled. The physician reads the X-ray and decides if the test is normal (no markers are seen at day 5) or abnormal (markers are present) and, depending on where the markers are clustered, if the constipation is a result of a generally sluggish colon (slow transit), a poor coordination and tone of the pelvic floor and the anal canal (pelvic floor dyssynergia or evacuation disorder), or a combination of both. Depending on the results, the physician can initiate and adjust therapy. For an accurate test, the patient abstains from using any laxatives during the study.

There is good diagnostic agreement between the radiopaque marker study and wireless motility capsule (WMC) in patients with chronic constipation [7].

CT of Abdomen and Pelvis (CTAP)

Computed tomography of the abdomen and pelvis (CTAP) is a diagnostic imaging test that helps in detecting diseases of the stomach, small bowel, colon, and other internal organs. It is often used to determine the cause of unexplained abdominal pain. CT scanning is fast, painless, noninvasive, and accurate, during which several X-ray beams and electronic X-ray detectors rotate around the patient's body measuring the amount of radiation being absorbed throughout the body. A special computer program then processes a large volume of data to create two-dimensional cross-sectional images of the body and then displays the images (thin slices) on a computer monitor. When the computer software reassembles these image slices, the result is a very detailed multidimensional view of the abdomen and pelvis. Modern CT scanners can image large sections of the body in just a few seconds, which is especially beneficial for the elderly and critically ill who cannot easily stay still to obtain images. The image below depicts a lateral view of the abdomen of a patient with PD suffering from severe constipation and fecal impaction that gives a honeycombing appearance of black and white dots, representing air and stool, respectively. Depending on the indication, CTAP may be performed with or without intravenous or intraluminal contrast (Figs. 4.18, 4.19, 4.20).

In this anteroposterior view of CTAP, there is extensive colonic diverticulosis, particularly in the sigmoid colon. There is thickening of the sigmoid wall, soft tissue stranding, and fluid in the sigmoid mesentery consistent with acute diverticulitis. The patient presented with history of long-standing constipation, mild left lower abdominal discomfort, and recent worsening of symptoms (Fig. 4.21).

Endoscopic Appearance at Colonoscopy in This Patient: The instrument could not be advanced through the sigmoid colon because of swelling and fixation of the bowel. A course of oral antibiotics (ciprofloxacin 500 mg by mouth twice daily plus metronidazole 250 mg by mouth thrice daily for 2 weeks) together with low fiber diet and oral osmotic laxatives led to clinical improvement.

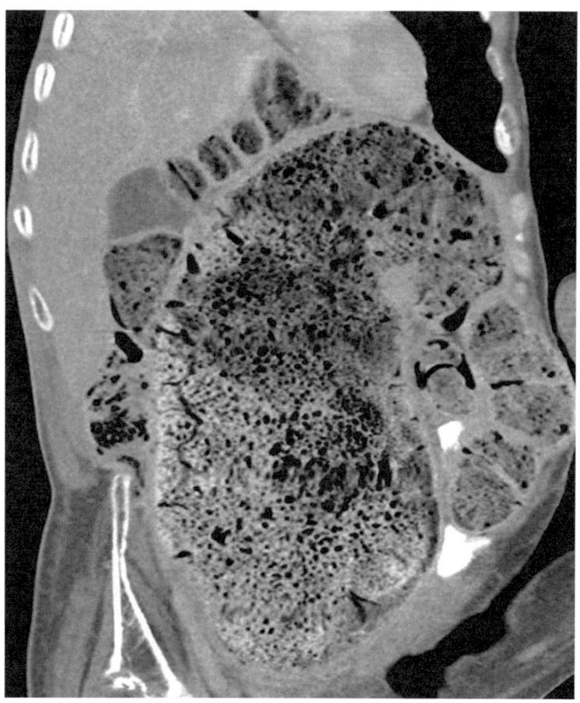

Fig. 4.18 Lateral view of the abdomen on abdominal and pelvic CT (CTAP) showing a honeycombing appearance of black and white dots, representing air and stool, respectively

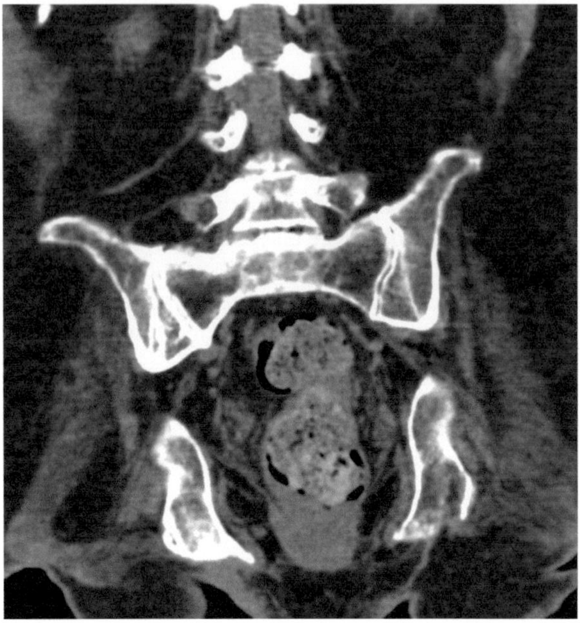

Fig. 4.19 Anteroposterior view of abdomino-pelvic CT showing fecal impaction

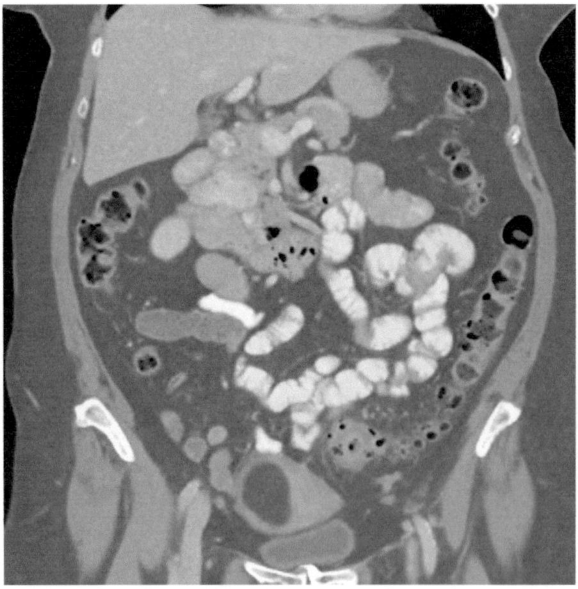

Fig. 4.20 Anteroposterior view of CTAP, showing extensive colonic diverticulosis, particularly in the sigmoid colon. There is thickening of the sigmoid wall, soft tissue stranding, and fluid in the sigmoid mesentery consistent with acute diverticulitis

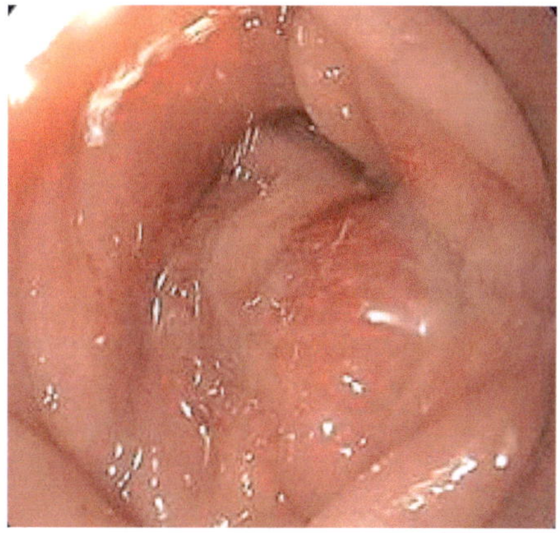

Fig. 4.21 Colonoscopy in acute diverticulitis. The instrument could not be advanced through the sigmoid colon because of swelling and fixation of the bowel

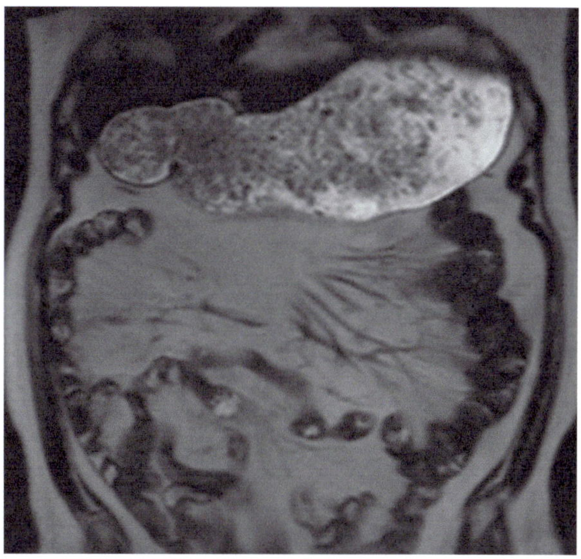

Fig. 4.22 MRI of abdomen revealing significant food and fluid retention in the stomach in this patient with gastroparesis secondary to PD

MRI Abdomen and Pelvis

An abdominal magnetic resonance imaging (MRI) scan is a test that uses powerful magnets and radio waves to create pictures of the inside of the abdomen and pelvis, without using radiation. Like a CT scan, it produces images, slicing the abdomen and pelvis reconstructing the anatomy of an organ or region. In the example below, there is significant food and fluid retention in the stomach in this patient with gastroparesis secondary to PD (Fig. 4.22).

Abdominal MRI in a patient with PD and chronic postprandial bloating in the upper abdomen with associated early satiety demonstrates gastric stagnation due to underlying gastroparesis.

MR Enterography (MRE)

This procedure uses MRI technology to obtain images of the small bowel. The procedure is painless, and there are no known risks, provided the patient has no metal in or on their body and is not pregnant. In addition to regular preparation for an MRI exam, prior to MRE, the patient is given two bottles of a special liquid to drink that serves to distend the bowel and marks the bowel for clear identification during the imaging study. Toward the end of the exam, the patient is given a small dose of glucagon, followed by an injection of gadolinium (an MRI contrast agent). Glucagon prevents the bowel from moving for a short time, which improves the quality of the images.

MRI Defecography

In general, MR imaging of the pelvis can characterize and differentiate benign from malignant masses, and it is useful in the pretreatment staging of cancer and the evaluation of benign pelvic floor processes. Dynamic pelvic floor MR imaging, or MR defecography with rectal contrast, can evaluate all pelvic compartments (anterior, middle, and posterior), muscles, ligaments, and fascia and is particularly useful in patients with suspected defecatory dysfunction or pelvic organ prolapse with or without urinary dysfunction. The study should be performed in at least a 1.5 T closed MR imaging unit with the patient supine and knees elevated, using 120–250 mL of ultrasound gel for rectal contrast and vaginal gel as appropriate. The urinary bladder should be partially distended. The study involves the acquisition of high-resolution T2WI through the pelvis in the coronal, axial, and sagittal planes in addition to dynamic sagittal sequences during squeezing (Kegel), straining (Valsalva), and evacuation (defecation) maneuvers (Fig. 4.23).

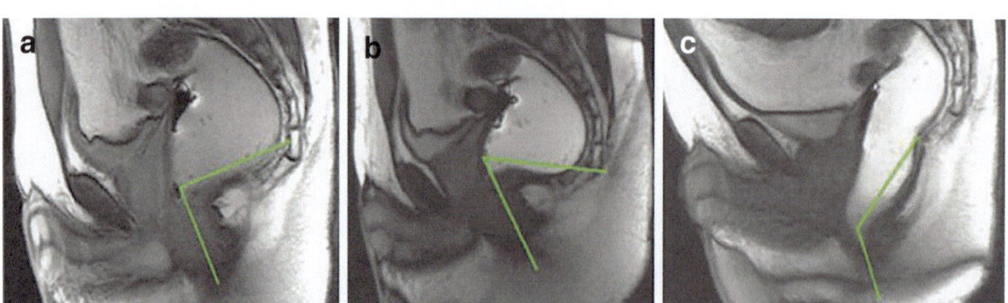

Fig. 4.23 MR defecography showing dynamic sagittal sequences during squeezing (Kegel), straining (Valsalva), and evacuation (defecation) maneuvers

Abdominal Ultrasound

This new point-of-care technique involves a 120-min ultrasonography examination using a 500-mL semiliquid test meal. The examiner may assess the antral contraction amplitude (ACA), the antrum contraction frequency (ACF), the motility index (MI), and the gastric antral cross-sectional area (CSA) at six time points: fasting for 12 h (TO), immediately after drinking the semiliquid test meal (T1), and at 30 (T30), 60 (T60), 90 (T90), and 120 (T120) min. We calculated the gastric emptying rate (GER) at different time points by using 87 the CSA. A study of patients with PD compared to healthy controls examined the GER between the groups and evaluated the correlation between the GER and gastrointestinal symptoms and motor symptoms of PD. According to the authors, gastric-filling ultrasonography is simple, reproducible, nonradioactive, and accurate and can distinguish the early gastric motility disorder of PD [8].

Gastric Alimetry (GA)

Gastric Alimetry (GA; Alimetry, Auckland, New Zealand) is a new test of gastric function that combines noninvasive body surface gastric mapping (BSGM) with concurrent validated symptom profiling. GA enables high-resolution characterization of gastric myoelectrical activity, being optimized to maximally separate the weak gastric signals from noise and to achieve a critical advance over conventional electrogastrography. The technique has been validated against antroduodenal manometry and invasive gastric electrical recordings and can identify specific patient subgroups with gastric neuromuscular dysfunction, including with symptom correlations. The merits of this new technology have not been shown in gut PD [9]. It is plausible that GA and GET will play a complementary role in clinical practice.

Gastric Emptying Study (GES, Scintigraphy)

The best way to assess for gastroparesis is the measurement of the gastric (stomach) emptying time over 4 h, after ingestion of a meal cooked with one egg and labeled with a small amount (1 mCi) of the radioactive technetium-99m. The patient ingests the standard egg sandwich meal over 10 min. Images of the abdomen are then taken in the front and back positions immediately after completion of the meal and at 30 min (normally, <96% retention), 1 h (normally <90% retention), 2 h (normally <60%), 3 h (normally <30%), and 4 h (normally <10% retention).

In the below case shown, most of the radioactivity (in black) is retained and outlining the stomach in both the front (left) and back (right) projections, with very little escaping in the esophagus and small bowel. Specifically, stomach % retention was 96%, at 1 h 93%, at 2 h 88%, at 3 h 74%, and at 4 h 64%, suggesting significant delay in stomach emptying. This patient with PD not only had frequent symptoms

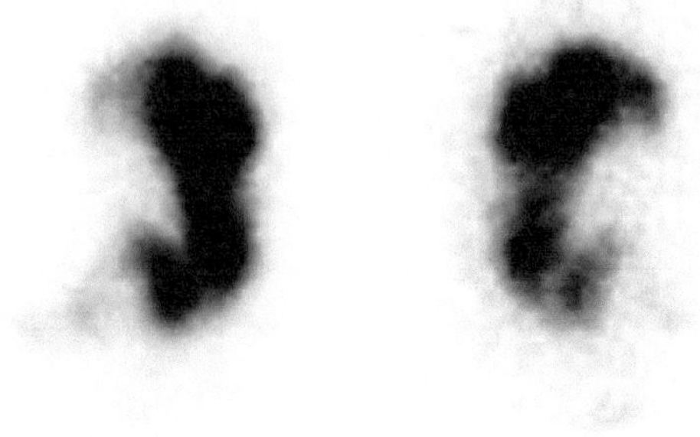

Fig. 4.24 Delayed gastric emptying scan, revealing 64% retention of the radioactive meal at 4 hours, suggestive of gastroparesis

of fullness after meals and nausea but also showed unpredictable response to dopaminergic drugs because of their delayed entry into the bowel and erratic absorption (Fig. 4.24).

Functional Assessment

High-Resolution Esophageal Manometry (HREM)

High-resolution esophageal manometry is a noninvasive test that shows whether the esophagus is working properly, contracting, and pushing food into the stomach with each swallow. This test measures the coordination of contractions and relaxations of the various esophageal muscles as they move food toward the stomach. This is a pivotal test to diagnose underlying esophageal motor function in patients with PD with suspected esophageal dysmotility (i.e., ineffective esophageal motility, hypercontractile esophagus), since it assesses intraluminal esophageal pressures, peristalsis (propulsion), and food bolus transit. During esophageal manometry, a thin, flexible, straw-like tube that contains pressure sensors is passed through the nose, down into the stomach. Using sips of water to swallow, the test measures how well the sphincter muscles at the top and bottom of the esophagus open and close, as well as the pressures, speed, and pattern of the wave of esophageal muscle contractions that move food along. The test is mostly useful in patients with difficult or painful swallowing. When studied, some patients with PD show multiple, forceful (spastic), poorly coordinated muscle contractions of esophagus (hypercontractile esophagus). Sometimes the contractions are weak and ineffective in the propulsion of a food or liquid bolus (ineffective esophageal motility). Other times, the esophageal sphincter muscles do not relax properly to let food enter the stomach (outflow obstruction), causing difficulty swallowing and regurgitation of food back up into the throat (esophagogastric outflow obstruction, or EGJOO). In one study of patients with PD presenting with symptoms of possible esophageal dysfunction using HR-EM, the frequency of these abnormalities was 55% for ineffective esophageal motility (IEM), 48% for esophageal spasm, and 39% for EGJOO. Only 6% of patients with PD in this cohort had normal esophageal motility [10] (Fig. 4.25).

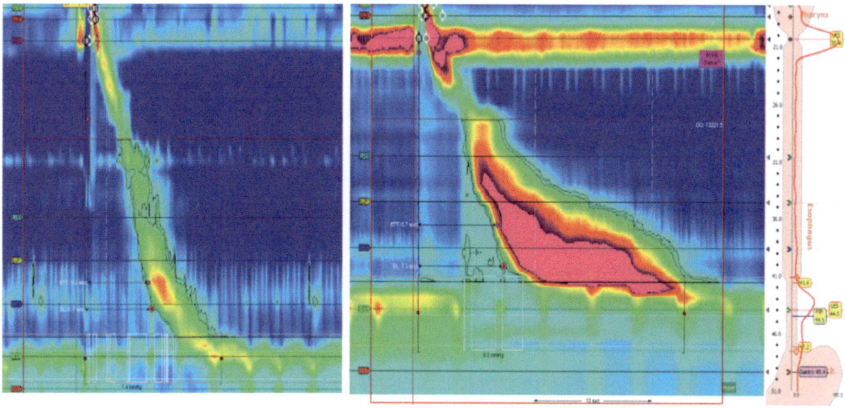

Fig. 4.25 Representative high-resolution esophageal manometry pressure tracings: *Left panel*: Normal pressure profile during simulated swallow of water in a normal subject. *Left panel*: Excessive contractility of the esophageal muscle during simulated swallow, suggestive of distal esophageal spasm (DES). Shown are the horizontal patterns of the upper esophageal (top) and lower esophageal sphincter (bottom) pressures recorded by the sensors. The diagonal pattern on the left and right panels shows the propulsion wave upon water swallowing (deep blue, low pressure; green, intermediate pressure; red, high pressure) as recorded by the sensors

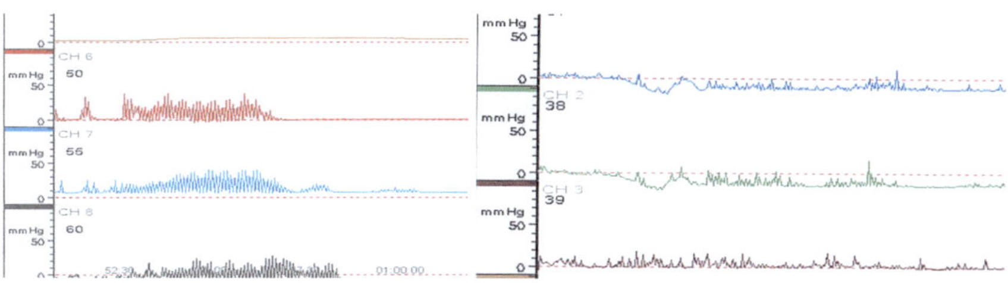

Fig. 4.26 Antroduodenal Manometry (ADM) recording. Left panel: Normal phase III of the migrating motor complex (MMC). Right panel: reduction in amplitude of contractions, consistent with intestinal myopathy

Antroduodenal Manometry (ADM)

This specialized manometry is not widely used or standardized, but it may help differentiate a myopathic from a neuropathic foregut process (i.e., visceral or autonomic neuropathy). Myopathic disorders are typically associated with low amplitude contractions (mean <40 mmHg in antrum), whereas in neuropathic disorders the amplitude of contractions is typically normal, but there is disorganization of the contractile response to a simulated meal with associated persistence of fasting motor patterns after meal ingestion (Fig. 4.26).

High-Resolution Anorectal Manometry (HR-ARM), Rectal Sensation Test (RST), and Balloon Expulsion Test (BET)

Intrarectal solid-state flexible probes can measure anal sphincter pressure, rectal compliance and sensation, and the recto-anal inhibitory reflex (RAIR). Anal sphincter function is assessed by measuring the resting, internal anal sphincter (IAS) pressure, squeeze external anal sphincter (EAS) pressure, and the anal canal length. Rectal balloon distention induces an early increase in rectal pressure, followed by a transient increase in anal pressure due to EAS contraction, followed by IAS relaxation. The RAIR may be absent in cases of muscle or nerve injury due to surgery (i.e., lower anterior colon resection) or pelvic radiation. Rectal sensation is evaluated by measuring the lowest perceived volume of balloon distention, desire to defecate, and the maximum tolerable volume. Manometry also assesses changes in anal and rectal pressures during attempted defecation. Low tone of IAS (hypotonicity) and shortened anal sphincter length may contribute to patients' urgency, frequency, and occasional fecal incontinence after pelvic radiotherapy; 75% of such patients exhibit one or more abnormal manometric parameters for sensory or motor functions, either isolated or in combination, while the RAIR may be absent or abnormal in up to 40% of patients [11].

For this, patients are instructed to give themselves one Fleet enema 2 h before the study. The procedure is done with the patient in the left lateral position with the patients' knees and hips flexed. Without anesthesia, a lubricated special, flexible, straw-sized catheter, mounted with pressure sensors and with an inflatable balloon at its tip, is then inserted into the rectum, through the anal canal. Rectal propulsive pressures are captured from the proximal sensor located in the rectal balloon, and the sensor catheter measures the anal pressures. With the catheter in place, patients are instructed to relax, squeeze, bear down, and cough. Resting and 20-s squeeze pressures are measured three times with 1 min of rest in between. Patients are then instructed to push and bear down to simulate evacuation for 20 s (with a deflated rectal balloon). The recto-anal inhibitory and cough reflexes are assessed. Rectal sensation testing is performed by distending the rectal balloon in 10 mL increments until the first sensory threshold was reached and in 30 mL increments thereafter until a maximum volume of 350 mL is reached or until the patient reports severe urgency (whichever is reached first). The balloon expulsion test is done using the balloon filled with 50 mL of water. Patients are asked to expel the balloon. If patients cannot expel the balloon in 2 min, it is deflated and removed manually. All pressure measurements and analyses are done using special software [12].

The pressure tracing below shows anorectal manometry patterns during simulated defecation in healthy patients and in patients with PD and associated dyssynergic defecation (less pressure in green anal bar at the bottom). Normally (left panel), increased rectal contraction (top green bar on blue background) is associated with normal anal canal relaxation (less pressure in green). In contrast, rectal contraction (top green bar on blue background) is associated with paradoxical anal contraction (yellow/red on green background) compromising fecal expulsion. To the right of each panel, there is a schematic view of the anorectal manometry catheter rectal sensors showing the rectal pressures (top) and anal sensors showing the anal canal pressures (bottom) (Fig. 4.27).

Figure shows two high-resolution anorectal manometry pressure tracings. A schematic view of the position of the anorectal manometry catheter sensors is shown on the right of each image. Shown are the rectal (top) and anal (bottom) pressures recorded by the sensors (deep blue, low pressure; green, intermediate pressure; red, high pressure). A balloon is stationed in the rectum. (A) Normal anal relaxation during simulated defecation in a normal subject. (B) Paradoxical anal contraction during simulated defecation, consistent with dyssynergic (poorly coordinated) defecation in a patient with PD.

There is high prevalence of anorectal disorders in PD. Using the newly devised London Classification for the assessment of HRM-ARM, abnormal rectal expulsion, incoordination (dyssynergia), and low contractility of the anal sphincter are the most common findings. Low anal sphincter pressures are significantly more common in women, probably a reflection of prior obstetrical injury. Poor rectal sensation is seen in nearly one third of PD patients and comparable among men and women [13]. Figure shows a three-dimensional rendering of the anorectal pressures recorded by the sensors (deep blue, low pressure; green, intermediate pressure; red, high pressure). (Left) Pressure profile during a squeeze effort. Maximum sphincter pressure 324 mmHg (normal <247 mmHg). There is contraction of the anal canal muscles, shown in red, during an attempt at expulsion of the rectal catheter during a high-resolution anorectal manometry. (Right) Pressure profile during bear-down (simulated attempted defecation). Residual anal pressure 165 mmHg (normal <65 mmHg) associated with impaired rectal expulsion (type II dyssynergia) (Fig. 4.28).

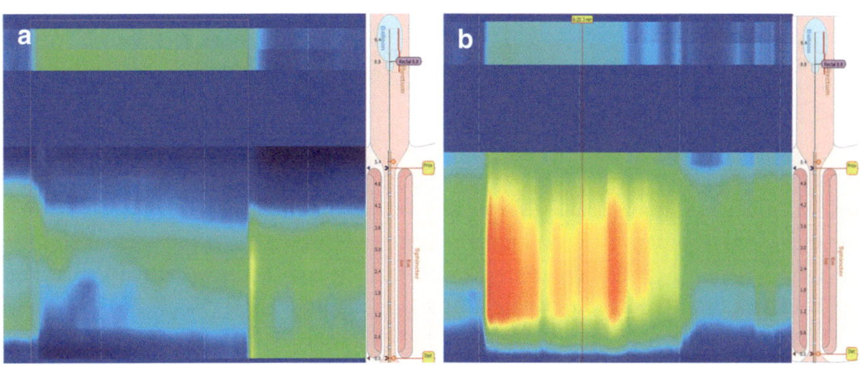

Fig. 4.27 High-resolution anorectal manometry: (**a**) Normal anal relaxation during simulated defecation in a normal subject. (**b**) Paradoxical anal contraction during simulated defecation, consistent with dyssynergic (poorly coordinated) defecation

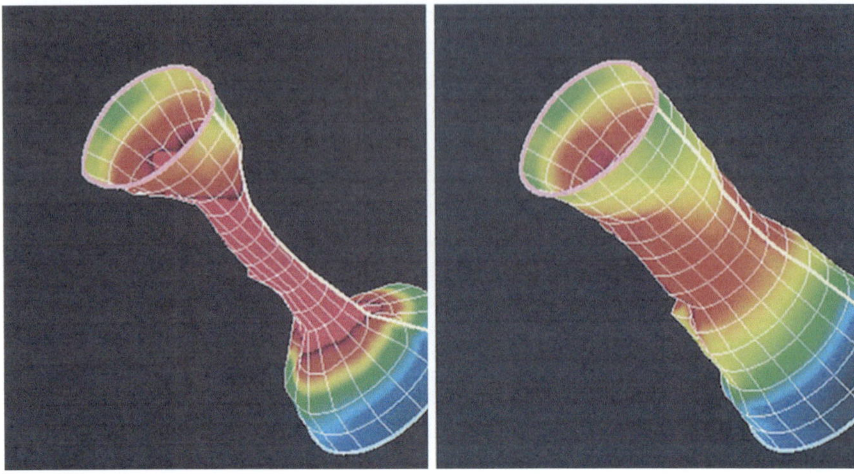

Fig. 4.28 Three-dimensional high-resolution anorectal manometry recording. (Right) Pressure profile during bear-down (simulated attempted defecation). Residual anal pressure 165 mmHg (normal <65 mmHg) associated with impaired rectal expulsion (type II dyssynergia)

Types of Defecatory Dyssynergia in PD

Normal: Adequate rectal propulsion and adequate anal relaxation.

Type I: Adequate rectal propulsion and paradoxical anal contraction

Type II: Impaired rectal propulsion and paradoxical anal contraction

Type III: Adequate rectal propulsion and incomplete anal relaxation

Type IV: Impaired rectal propulsion with incomplete anal relaxation

Impaired or inadequate rectal propulsion is defined by HR-ARM as rectal pressure <41 mmHg in men and <24 mmHg in women. Normal anal relaxation during attempted defecation is defined as a decrease in pressure of ≥20%. (Fig. 4.29)

Complimentary Tests in Chronic Constipation, Stool Impaction, and Fecal Incontinence

A full assessment involves the use of all three tests: BET, balloon expulsion test; RST, rectal sensation test; and HR-ARM, high-resolution anorectal manometry. All three tests may be performed in tandem, as part of one visit to a specialized anorectal motility unit. MR defecography (performed in the radiology department) and BET are important tools to evaluate evacuation disorder (EV) (Fig. 4.30).

Figure below depicts an HR-ARM image (left) and an MR defecography (right) in a 65-year-old woman with PD who presented with episodes of debilitating diarrhea fecal incontinence interspersed with periods of constipation.

HR-ARM (left) showed low resting and squeeze anal pressures. The squeeze duration was adequate. During the push maneuver, the residual anal pressure was high, and the percent (%) is only 6% (normal 15–35%). The recto-anal inhibitory

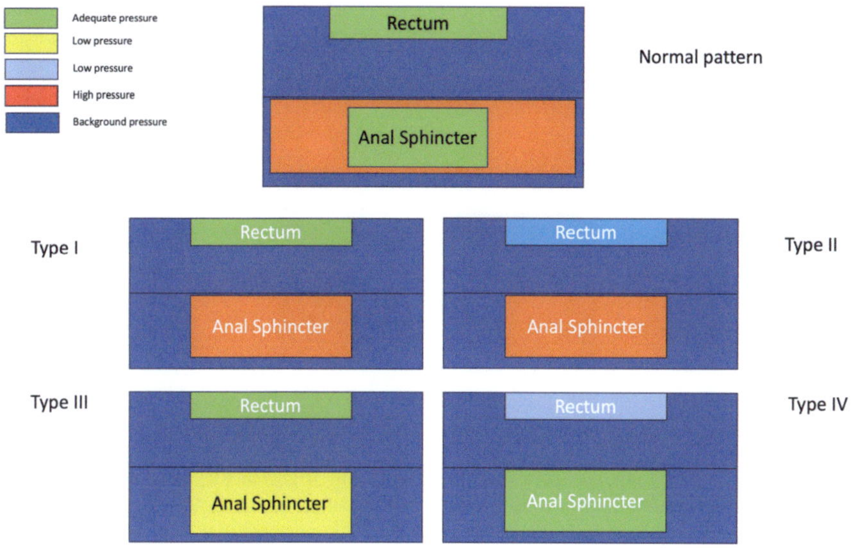

Fig. 4.29 Rectal and anal pressure profiles normally and with the 4 types of dyssynergia

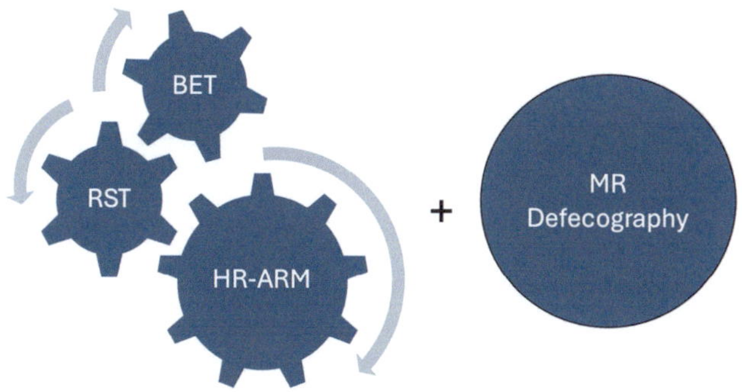

Fig. 4.30 High-resolution anorectal manometry (HR-ARM), Rectal Sensory Testing (RST), Balloon Expulsion Testing (BET), and MR Defecography are complimentary tests in the evaluation of patients with chronic constipation, stool impaction, and fecal incontinence

reflex (RAIR) was not seen. Rectal sensation was mildly impaired. Balloon expulsion was not achieved.

MR defecography (right) at rest showed a pelvic hiatus 6.4 cm ("H"-line; normal ≤5 cm) with pelvic floor descent 2.7 cm ("M"-line; normal ≤2 cm): 2.7 cm. The anorectal angle was 106° (normal). Upon squeeze there was normal pelvic floor ascent. Upon evacuation, the inferior margin of the bladder extended 1 cm below the pubococcygeal line (anterior compartment): greater than expected descent of the vagina (middle compartment). There was no enterocele, peritoneocele, or rectal intussusception. There was normal appearance of the levators and puborectalis. There was poor evacuation (approximately one third of contrast material was

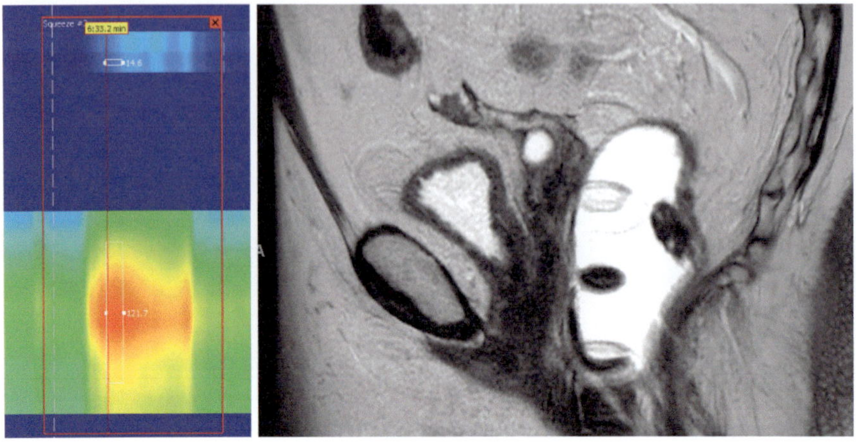

Fig. 4.31 HR-ARM image (left) and an MR defecography (right) in a 65-year-old woman with PD. The HR-ARM (left) shows low resting and squeeze anal pressures, while defecography shows mild widening of the pelvic hiatus in the resting state, with intact squeeze response and tri-compartmental descent

Fig. 4.32 Outline of the balancing forces behind a normal defecation

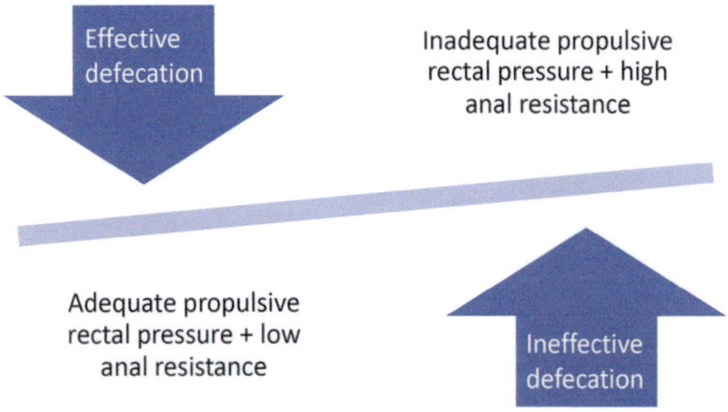

evacuated). Overall impression was mild widening of the pelvic hiatus in resting state with intact squeeze response and tricompartmental descent (Fig. 4.31).

Diagram below outlines the balancing forces behind a normal defecation. Alterations studied by HR-ARM facilitate the management of patients with PD and constipation, urgency, fecal impaction, and, at times, fecal incontinence (Fig. 4.32).

Endoluminal Functional Lumen Imaging Probe (EndoFLIP)

Impedance Planimetry Using Endoluminal Functional Lumen Imaging Probe (EndoFLIP)

The functional luminal imaging probe (FLIP) uses high-resolution planimetry to provide a three-dimensional image of the esophageal lumen by measuring diameter, volume, and pressure changes [14]. The EndoFLIP catheter includes a distensible

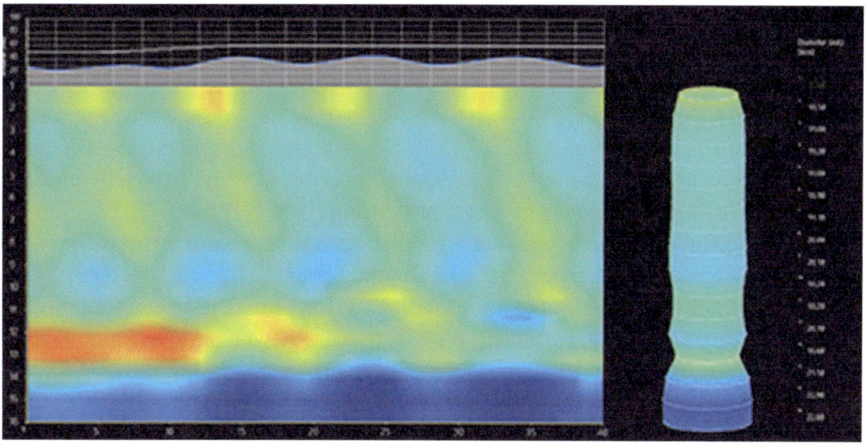

Fig. 4.33 Representative FLIP topography image from a 72-year-old man with intermittent dysphagia and long-standing history of PD. The study was done to exclude any esophageal dysmotility and EGJOO as potential causes of dysphagia. Patient could not tolerate high-resolution esophageal manometry experiencing recurrent spells of coughing and nasal regurgitation during water swallows. FLIP topography was normal, with adequate distensibility of the EGJ (right) and normal recurrent antegrade contractile activity (left). The use of EndoFLIP in this case obviated the need for another attempt for HREM and its associated patient discomfort

balloon encasing many pairs of sensors and a single distal pressure sensor. It is placed through the mouth during an upper endoscopy, such that the balloon traverses the area of interest, most often the junction of the esophagus with the stomach. In contrast, to HREM, EndoFLIP is performed while the patient is receiving intravenous propofol anesthesia and is generally well tolerated. The balloon is then distended with a conductive fluid through a mechanical pump while the FLIP system converts the readings into a three-dimensional rendering of the esophageal lumen in real time and measures the relationship between the cross-sectional area (mm^2) and the distention pressure to generate a luminal distensibility. FLIP topography converts the readings to color-coded luminal diameter plots over time allowing the evaluation of distensibility across the EGJ, as well as visualizing secondary peristalsis—the esophageal contractile response to distension in the esophageal body. As such, it may provide some information on the state of esophageal body contractility as well. A normal EGJ distensibility index (EGJ-DI) is >2.8 mm^2/mmHg, and normal EGJ diameter is >13 mm. Further, the presence of repetitive antegrade contractions is considered a normal response to distension. Reduced EGJ-DI and/or diameter is often seen in patients with disorders of EGJ outflow, such as achalasia or EGJ outflow obstruction (EGJOO). EGJ-DI of <2 mm^2/mmHg is abnormal, whereas EGJ diameter of <13 mm is likely abnormal and can serve as a supportive measure, when EGJ-DI is indeterminate (2–3 mm^2/mmHg) [15] (Fig. 4.33).

24-Hour Impedance-pH Monitor

A 20-h esophageal impedance-pH monitoring is a key diagnostic tool to evaluate patients with PD who suffer from heartburn, chest pain, and regurgitation and they have no abnormalities on endoscopy. The spaghetti size, wire-like sensor is typically placed 5 cm above the lower esophageal junction with the stomach, and it is connected to a portable digital recorder that stores pH data for up to 24 h. Positioning of the catheter may also be based on the pH differential between the two pH sensors (one in the stomach, 10 cm below the LES, and a proximal one in the esophagus, 5 cm above the LES). A plain chest radiograph may be obtained in cases where the location of the catheter is uncertain. Typically, in patients who are not receiving acid blocker therapy, the stomach sensor reveals low pH (~1.0), while the esophageal sensor shows higher pH (>4.0). Patients are instructed to keep a diary of symptoms, meal intake times, and body position, such as supine position at night, regardless of sleep (nighttime supine period). Acid suppressive drugs, typically proton pump inhibitors (PPI), may or may not be used during the study, depending on the reason for the pH monitoring (diagnosis, response to therapy, symptom persistence). The parallel impedance monitoring helps identify nonacid reflux in many cases of normal pH monitoring. Alternatively, continuous gastric pH monitoring may be performed to validate and quantify low acid production in the stomach (gastric hypochlorhydria or achlorhydria), a result of long-standing *H. pylori*-related atrophic gastritis, which may affect levodopa absorption. The study may be challenging to do in patients with PD who are troubled by oropharyngeal dysphagia and nasal regurgitation who may not easily tolerate the placement and retention of the pH probe. In such cases, 96-h wireless esophageal pH-metry is preferable (Fig. 4.34).

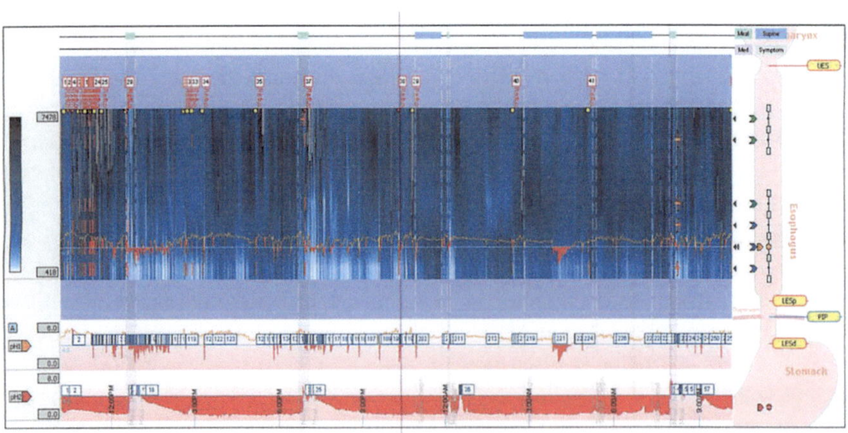

Fig. 4.34 The pH tracing of the distal esophagus sensor (top red line) shows moderately abnormal acid reflux (% pH <4.0 of 12.8; normal <6), mostly in the upright position, while the stomach sensor (bottom red line) shows normal stomach acidity (% pH <4.0 of 96%), throughout the 24-h period, evidence against lack of gastric acidity

96-Hour Wireless Esophageal pH Monitoring (pH-metry)

Ambulatory esophageal pH monitoring is a diagnostic tool in patients with PD suffering by heartburn and regurgitation. During an endoscopy, the 96-h wireless esophageal pH recording capsule is attached to the mucosa, 6 cm above the junction between the esophagus and stomach (EGJ). The subsequent wireless pH-metry is better tolerated, allowing patients to eat freely and thereby achieving more real-life data and for longer recording period (96 h) which allows for measurement of the effect of diet or drugs on esophageal acid exposure. The study should be considered when trying to establish the presence of abnormal acid exposure as a potential cause for the patients' symptoms. It is particularly helpful in patients with PD, when catheter-based pH monitoring cannot be tolerated. Like the pH catheter system, patients are studied while on or off therapy with acid blockers, keeping record of their symptoms, meals, and position [16].

In most cases, esophageal ambulatory pH monitoring is performed "off" acid suppressive therapy using proton pump inhibitors (PPI) for a minimum of 10 days before and 4 days during the study. The wireless 96-h Bravo® pH system (Medtronic, Sunnyvale, CA) comprised of a delivery system, pH recording capsule with integrated radio transmitter, a radio receiver and recording device, and software for data analysis. Data are transmitted to a receiver worn on a strap or belt. Patients are instructed to document mealtimes, position change, and their symptoms. Reflux event detection is set at a sampling frequency of every 6 s and defined as two consecutive measurements of pH <4 (i.e., 12-s acid exposure). These data are downloaded and analyzed by proprietary software. Patients are instructed to carry out normal daily activities, consuming liberal diet (to induce reflux symptoms) during the first 48 h of the study, followed by a restricted diet (to curtail acid reflux) for the next 48 h, and to record their symptoms using the recorder button and a diary. There is usually no need to record the specifics of the meals (nature, caloric content, or timing). Esophageal acid exposure time (AET) is defined as the total time the pH in the distal esophagus is lower than 4.0, expressed as a percentage; values ≥6% per 24 h define an abnormal AET. For each patient, the average from the first 2 days (on liberal diet) is considered as the baseline AET, and it is compared with the average from the latter 2 days on restrictive diet (Figs. 4.35, 4.36).

Laryngo-Esophageal pH Monitoring (Dx–pH Measuring System)

This tool determines the impact of laryngopharyngeal reflux (LPR) on symptoms, such as difficulty clearing the throat, chronic cough, or aspiration in the lungs. Since it measures acid, weakly acidic gastroesophageal reflux (WAGR), and alkaline reflux, the Dx–pH is very useful in the evaluation of not only patients with possible acid reflux but also potential bile refluxers, with alkaline exposure in the pharynx, larynx, and proximal esophagus (Fig. 4.37).

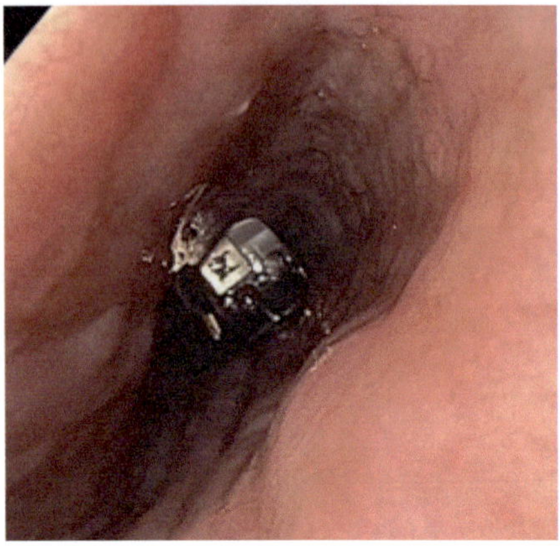

Fig. 4.35 Endoscopic image of the Bravo pH capsule placed in the distal esophagus and recording esophageal acid exposure over the ensuing 96 h

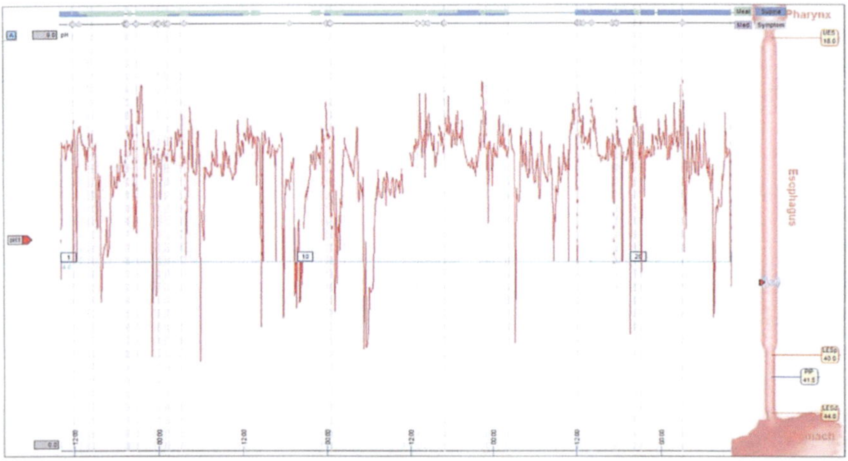

Fig. 4.36 In this 96-h wireless esophageal pH recording, the *Y* axis shows the distal esophageal pH (range 0–8.0), while the *X* axis shows the recording time (0–96 h). While most of the graph shows the pH to be 5.0–6.0, there are short-lived episodes where the pH falls below the 4.0 line (in blue). These events happened mostly during the first 48 h of the study, during which period the patient was on a liberal diet, and they were less frequent during the latter 48 h of the study, while the patient was consuming a restrictive, anti-reflux diet

Wireless Motility Capsule (WMC, SmartPill®)

The WMC (SmartPill; Medtronic, Sunnyvale, California, USA) is an ambulatory, noninvasive, and nonradioactive diagnostic sensor that continuously samples intraluminal pH, temperature, and pressure as it moves through the gastrointestinal tract. This technology quantifies transit in all gut regions in a single test, and it has been increasingly used for the diagnosis of many gut symptoms seen in PD, such as

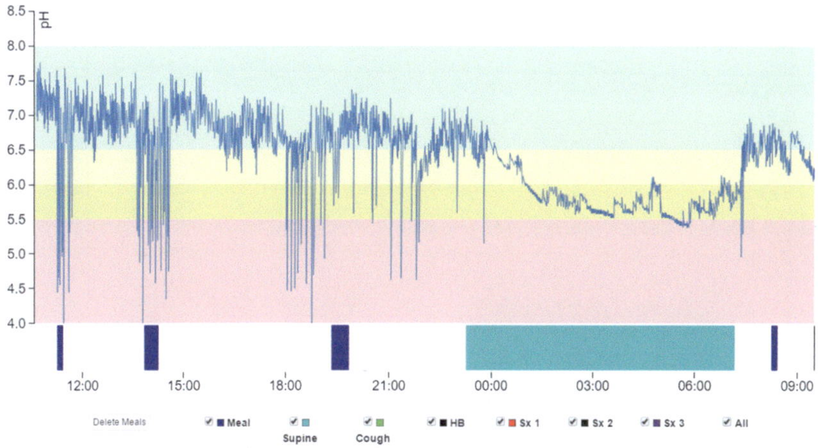

Fig. 4.37 Laryngo-Esophageal pH Monitoring revealing transient episodes of pharyngeal acid exposure

bloating, early postprandial fullness, nausea, and constipation, by measuring gastric emptying time, as well as small bowel and colonic transit times. The patient first ingests a meal consisting of a SmartBar (260 kcal, 2% fat, 2 g fiber) and 120 mL water to initiate the postprandial motility pattern after an overnight fast. After the meal, the patient swallows the capsule with 50 mL water. The capsule wirelessly connects to a data receiver, and the patient keeps a diary for recording bowel movements, food intake, sleep, and gut symptoms while continuing their regular diet and life routines. Physical restrictions include no strenuous activities, such as sit-ups, abdominal crunches, and prolonged aerobic activity (>15 min), which could affect pressure measurements. Additionally, patients refrain from the use of gastrointestinal medications that could affect motility (i.e., laxatives) or gastric pH (i.e., proton pump inhibitors). Patients are asked to fast for 6 h after capsule ingestion, after which they ingest a regular meal. This meal would allow for the evaluation of the postprandial response, which is the change in contractile activity of the small bowel. Patients are then instructed to continue their regular diet and routine. After 5 days, patients return the data receiver and the diary, and the downloaded data are analyzed using special display software.

In this representative case of a 74-year-old man with PD, the gastric emptying time is 2 h, 10 min (normal <4 h); the small intestinal transit time is 3 h, 15 min (normal 2.5–6 h); and the colon transit time is 102 h, 18 min (normal <59 h). The tracing shows the temperature (in blue), the luminal pH (in green), and the intraluminal pressures (in red). The combined values of these three parameters define the location of the capsule and, in turn, its location (stomach, small intestine, and colon) as it traverses the gut. Note the gradually increasing luminal pH as the sensor is moving into the colon at 70 min and the increased, dense red spike activity between 82 and 98 h prior to passage of the capsule through the anal canal (Fig. 4.38).

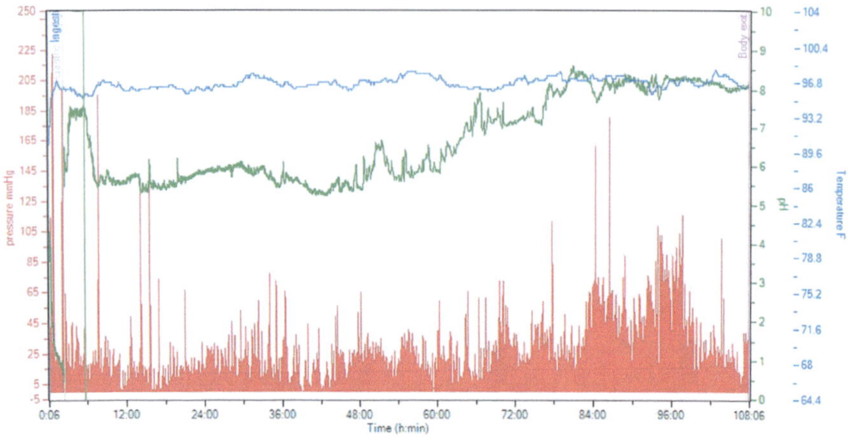

Fig. 4.38 Wireless motility capsule (WMC) tracing showing the temperature (in blue), the luminal pH (in green), and the intraluminal pressures (in red). Collectively, these variables reflect gastric, small intestinal, and colon transit times

Electrogastrography (EGG)

Patients with PD often present with gastric symptoms. Electrogastrography (EGG) can noninvasively assess gastric electric activity and may be useful in early PD diagnosis [17]. The technique involves placement of surface electrodes, on the anterior abdominal wall, to record the electrical control activity of the stomach, which normally occurs at 3 cpm (cycles per minute). Altered frequency (bradygastria, tachygastria, or mixed) and reduction in the amplitude of the postprandial electrical gastric signal are seen in patients with gastroparesis and nausea. Such dysrhythmias may be seen in patients with dyspepsia, with or without evidence of gastric emptying delay. EGG has not been adequately explored in the clinical evaluation and management of patients with PD, but it could provide an easy management tool. Early and untreated PD patients may show irregular EGG waves and high preprandial instability coefficient of dominant frequency (ICDF), an index of EGG irregularity, and better accuracy than the olfactory test for early PD diagnosis [18].

This noninvasive technique involves placement of surface electrodes, placed on the anterior abdominal wall, to record the electrical control activity of the stomach, which normally occurs at 3 cpm (cycles per minute). Altered frequency (bradygastria, tachygastria, or mixed) and reduction in the amplitude of the postprandial electrical gastric signal are seen in patients with gastroparesis and nausea. Such dysrhythmias may be seen in patients with dyspepsia, with or without evidence of gastric emptying delay. EGG has not been adequately explored in the clinical evaluation and management of patients with PD, but it could provide an easy management tool. In the recording below, the postprandial electrical activity of the small intestine (8–12 cpm) and the colon (18–20 cpm) can also be seen.

^{13}C-Octanoate Breath Test

This test of gastric emptying consists of a scrambled egg where the yolk has been mixed with 91 mg 13C-octanoic acid, a stable, nonradioactive isotope. The egg is eaten with two slices of white bread and 15 g of margarine, immediately followed by 150 mL of water. Before the test, a breath sample is taken for baseline measurements [19].

Glucose and Lactulose Breath Tests

In these tests, glucose or lactulose, a nonabsorbable sugar solution, is swallowed, and hydrogen and methane gases are measured in the exhaled breath with sampling every 15 min. Gas levels rise as the sugar solution passes through the intestine, resulting from fermentation of the sugar by bacteria. If the gas levels rise quickly and to an abnormal level, a diagnosis of small intestinal bacterial overgrowth (SIBO) can be made and treated. Left panel: Positive test for hydrogen, diagnostic of SIBO. There is a slow rise in the exhaled hydrogen over 1 h, starting at 60 min, and slowly declining, over 180 min of testing. Right panel: Negative test for methane release. There is no evidence of methane exhaled in the breath over the 180-min period. The presence of these two gases in the exhaled breath either together or in isolation dictates differential treatment with antibiotics (Fig. 4.39).

Sialometry

The amount of saliva flow varies considerably from individual to individual and correlates poorly with the sensation of a dry mouth. Sialometry is a measure of saliva flow. Many different techniques have been devised, yet none are perfect. Four major salivary glands and many minor salivary glands generate saliva. The parotid glands produce saliva with stimulation, such as eating, sucking on a lemon candy, chewing gum, or smelling a tasty meal. The submandibular glands produce saliva

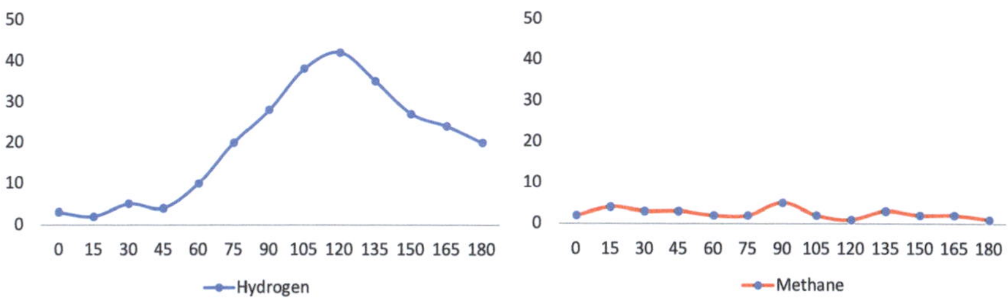

Fig. 4.39 Lactulose breath test. Left panel: Positive test for hydrogen, diagnostic of SIBO. There is a slow rise in the exhaled hydrogen over 1 h, starting at 60 min, and slowly declining, over 180 min of testing. Right panel: Negative test for methane release. There is no evidence of methane exhaled in the breath over the 180-min period

continuously, thus keeping the mouth moist even without eating. The following tests can measure saliva production, but they are rarely used in practice: (1) Unstimulated whole saliva production. Patient is asked to sit without talking or chewing and to spit any saliva that accumulates in the floor of the mouth into a pre-weighed tube. (2) Stimulated whole saliva production. Patient is asked to chew on a piece of pre-weighed gauze for a period of 5–15 min. The accumulated saliva is then measured, by spitting into a pre-weighed tube. (3) Stimulated parotid saliva production. Special suction cups are placed over the two small openings in the mouth where the saliva produced by the parotid glands is released. The saliva produced is collected in pre-weighed tubes for a period of 5–15 min. Lemon juice is usually applied to the tongue to stimulate parotid saliva flow.

Autonomic Function Tests

Autonomic nervous system testing helps to determine the presence, severity, and localization of autonomic dysfunction, and it involves the following components: (1) deep breathing, Valsalva, and tilt test with monitoring of heart rate, blood pressure (both continuous from finger and intermittent from arm), end tidal CO_2, respiratory frequency, and cerebral blood flow velocity from middle cerebral artery. (2) Skin biopsy for evaluation of epidermal and sweat gland nerve fiber density. (3) Sudomotor function evaluation using quantitative sudomotor axon reflex test (QSART) or electrochemical skin conductance (ESC). All these tests are performed in sophisticated laboratories and are rarely used in patients with PD unless they are severely symptomatic and troubled by syncope, chronic fatigue, low blood pressure, dizziness, postural tachycardia, abnormal sweating, bladder and bowel dysfunction, or concomitant diabetes with neuropathy. The tests are often used to monitor disease progression and response to treatment [20].

References

1. Triadafilopoulos G, Mashimo H, Tatum R, O'Clarke J, Hawn M. Mixed esophageal disease (MED): a new concept. Dig Dis Sci. 2023;68(9):3542–54. https://doi.org/10.1007/s10620-023-08008-x. Epub 2023 Jul 20.
2. Su A, Gandhy R, Barlow C, Triadafilopoulos G. A practical review of gastrointestinal manifestations in Parkinson's disease. Parkinsonism Relat Disord. 2017;39:17–26. https://doi.org/10.1016/j.parkreldis.2017.02.029. Epub 2017 Feb 23.
3. Rosa-E-Silva L, Gerson L, Davila M, et al. Clinical, radiologic, and manometric characteristics of chronic intestinal dysmotility: the Stanford experience. Clin Gastroenterol Hepatol. 2006;4:866–73. https://doi.org/10.1016/j.cgh.2006.05.001.
4. Dziewas R, Warnecke T, Labeit B, Claus I, Muhle P, Oelenberg S, Ahring S, Wüller C, Jung A, von Itter J, Suntrup-Krueger S. Systematic approach to contextualize findings of flexible endoscopic evaluation of swallowing in neurogenic dysphagia- towards an integrated FEES report. Neurol Res Pract. 2024;6(1):26. https://doi.org/10.1186/s42466-024-00321-8. PMID: 38720388; PMCID: PMC11080162.
5. Antunes C, Sloan JA. Esophageal radiography interpretation: a primer for the gastroenterologist. Curr Gastroenterol Rep. 2023;25(12):363–73. https://doi.org/10.1007/s11894-023-00903-7. Epub 2023 Nov 8.

6. Madigan KE, Smith JS, Evans JK, Clayton SB. Elevated average maximum intrabolus pressure on high-resolution manometry is associated with esophageal dysmotility and delayed esophageal emptying on timed barium esophagram. BMC Gastroenterol. 2022;22(1):74. https://doi.org/10.1186/s12876-022-02165-5. PMID: 35189822; PMCID: PMC8859877.
7. Rao SS, Coss-Adame E, Valestin J, Mysore K. Evaluation of constipation in older adults: radioopaque markers (ROMs) versus wireless motility capsule (WMC). Arch Gerontol Geriatr. 2012;55(2):289–94. https://doi.org/10.1016/j.archger.2012.04.003. Epub 2012 May 7.
8. Zou X, Chen X, Wen Y, Jing X, Luo M, Xin F, Tang Y, Hu M, Liu J, Xu F. Gastric-filling ultrasonography to evaluate gastric motility in patients with Parkinson's disease. Front Neurol. 2024;15:1294260.
9. Wang WJ, Foong D, Calder S, Schamberg G, Varghese C, Tack J, Xu W, Daker C, Carson D, Waite S, Hayes T, Du P, Abell TL, Parkman HP, Huang IH, Fernandes V, Andrews CN, Gharibans AA, Ho V, O'Grady G. Gastric alimetry expands patient phenotyping in gastroduodenal disorders compared with gastric emptying scintigraphy. Am J Gastroenterol. 2024;119(2):331–41.
10. Su A, Gandhy R, Barlow C, Triadafilopoulos G. Clinical and manometric characteristics of patients with Parkinson's disease and esophageal symptoms. Dis Esophagus. 2017;30(4):1–6. https://doi.org/10.1093/dote/dow038.
11. Bharucha AE, Basilisco G, Malcolm A, Lee TH, Hoy MB, Scott SM, Rao SSC. Review of the indications, methods, and clinical utility of anorectal manometry and the rectal balloon expulsion test. Neurogastroenterol Motil. 2022;34(9):e14335. https://doi.org/10.1111/nmo.14335. Epub 2022 Feb 27. PMID: 35220645; PMCID: PMC9418387.
12. Frye J, Rao SSC. Anorectal manometry: when, how to perform and interpret, and is it useful? Am J Gastroenterol. 2024;119(8):1449–55.
13. Zhou W, Triadafilopoulos G, Gurland B, Halawi H, Becker L, Garcia P, Nguyen L, Miglis M, Muppidi S, Sinn DI, Jaradeh S, Neshatian L. Differential findings on anorectal manometry in patients with Parkinson's disease and defecatory dysfunction. Mov Disord Clin Pract. 2023;10(7):1074–81. https://doi.org/10.1002/mdc3.13755. PMID: 37476327; PMCID: PMC10354598.
14. Donnan EN, Pandolfino JE. EndoFLIP in the esophagus: assessing sphincter function, wall stiffness, and motility to guide treatment. Gastroenterol Clin N Am. 2020;49(3):427–35. https://doi.org/10.1016/j.gtc.2020.04.002. Epub 2020 Jun 14. PMID: 32718562; PMCID: PMC7387752.
15. Ahuja NK, Clarke JO. The role of impedance planimetry in the evaluation of esophageal disorders. Curr Gastroenterol Rep. 2017;19(2):7. https://doi.org/10.1007/s11894-017-0544-2.
16. Triadafilopoulos G, Kamal A, Clarke JO. Value of 96-h ambulatory esophageal pH monitoring in the assessment of patients with refractory acid reflux symptoms and their response to anti-reflux diet. Dis Esophagus. 2023;36(7):doac102. https://doi.org/10.1093/dote/doac102.
17. Lu CL, Shan DE, Chen CY, Luo JC, Chang FY, Lee SD, Wu HC, Chen JD. Impaired gastric myoelectrical activity in patients with Parkinson's disease and effect of levodopa treatment. Dig Dis Sci. 2004;49:744–9.
18. Araki N, Yamanaka Y, Poudel A, Fujinuma Y, Katagiri A, Kuwabara S, Asahina M. Electrogastrography for diagnosis of early-stage Parkinson's disease. Parkinsonism Relat Disord. 2021;86:61–6. https://doi.org/10.1016/j.parkreldis.2021.03.016. Epub 2021 Mar 28.
19. von Gerichten J, Elnesr MH, Prollins JE, De Mel IA, Flanagan A, Johnston JD, Fielding BA, Short M. The [^{13}C] octanoic acid breath test for gastric emptying quantification: a focus on nutrition and modeling. Lipids. 2022;57(4–5):205–19. https://doi.org/10.1002/lipd.12352. Epub 2022 Jul 7. PMID: 35799422; PMCID: PMC9546385.
20. Jones PK, Gibbons CH. Autonomic function testing: an important diagnostic test for patients with syncope. Pract Neurol. 2015;15(5):346–51. https://doi.org/10.1136/practneurol-2015-001102. Epub 2015 Jun 24.

Infections in Gut PD

5

Helicobacter pylori Infection

Helicobacter pylori (*H. pylori*), a Gram-negative, spiral-shaped, microaerophilic, and flagellated bacterium, is widely prevalent worldwide, infecting approximately half of the world's population. The bacterium predominantly colonizes the gastric mucosa in humans and predisposes individuals to various gastric mucosal diseases, including gastritis, gastric ulcers, and gastric cancer [1]. With a prevalence range of 26.4–87.9% in PD patients, *H. pylori* infection has been noted to disrupt the delicate equilibrium of the microbiota within the gastrointestinal tract [2].

Helicobacter pylori is a germ that lives in the mucus lining of the stomach and may cause ulcers and stomach cancer. This infection may not cause symptoms, but it may be associated with a 45% increased risk of PD, since it triggers an immune response that might cause nerve damage. *H. pylori* may contribute to motor fluctuations in PD, since it may interfere with the absorption of levodopa with increased delay in onset of action of levodopa and decreased on-time when the motor symptoms are adequately controlled [3]. If identified, the infection should be treated aggressively and eradicated, not only to mitigate the chances of motor fluctuations, but also to reduce the risk for ulcers and their complications, such as bleeding, as well as stomach cancer.

Depending on age, sex, socioeconomic status, and geographical origin, almost half of the world's population carries *H. pylori* in their stomach. Although most infected people have no symptoms, these bacteria may cause persistent stomach inflammation (gastritis), but some may develop ulcers and even gastric cancer. It is unclear if *H. pylori* gastritis increases the risk of PD and if the resulting inflammation is the cause or effect of the digestive symptoms of PD. Yet, *H. pylori* infection interferes with L-dopa absorption, thereby affecting treatment efficacy and contributing to motor fluctuations. In turn, eradication of the infection improves L-dopa bioavailability, so patients without *H. pylori* experience a longer duration of the "on state" lasting for several months, whereas untreated patients or those who have failed eradication therapy have delayed, shorter, or off times [1]. However, despite

G. Triadafilopoulos, *The Gut in Parkinson's Disease*,
https://doi.org/10.1007/978-3-031-77171-2_5

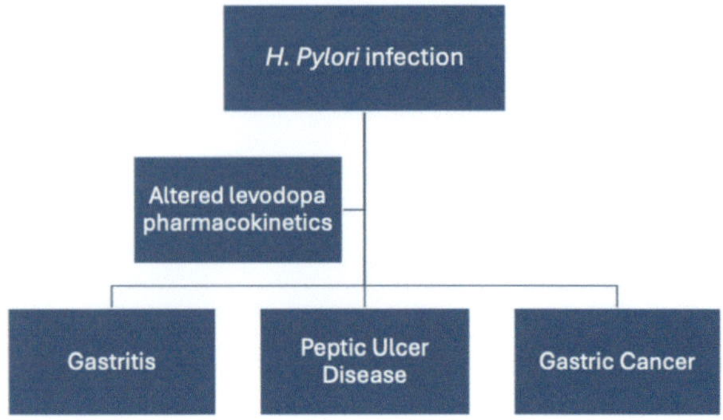

Fig. 5.1 Clinical implications of *Helicobacter pylori* infection in patients with PD

the improvement of motor fluctuations, *H. pylori* eradication has no impact on disease severity measured by standard questionnaires. Importantly, however, longstanding *H. pylori* infection may lead to changes in the gut microbiota secondary to stomach atrophy and decreased gastric acid secretion, which in turn impairs L-dopa absorption in the small intestine.

For these reasons, all patients with PD should be tested for *H. pylori* infection and treated as needed. For those who present with gastrointestinal symptoms, endoscopy with stomach biopsies is preferred, since it can detect an active *H. pylori* infection and quantify the severity of gastritis, which may affect levodopa absorption. For those patients without symptoms, noninvasive tests for active *H. pylori* infection include the urea breath test (impractical for patients with PD) and the stool antigen test, based on a stool sample. Blood antibody tests are easy and useful to detect immune response to the infection, but if they are positive, they cannot distinguish a past, previously treated, or presently active infection. Two months after a course of therapy, the stool antigen test should be used to confirm eradication [4] (Fig. 5.1).

The first-line treatment for the infection involves a 2-week course of "quadruple" therapy with a proton pump inhibitor (PPI), such as omeprazole (40 mg twice daily) or equivocal agent, amoxicillin (1 g twice daily), clarithromycin (500 mg twice daily), and bismuth (120 mg twice daily). This regiment is tough on the stomach of many patients, and a high degree of compliance is needed to ensure completion, thereby avoiding treatment failures and development of resistance. The "triple" therapy regimen (same as above, but without bismuth) may be more tolerable [5].

Small Intestinal Bacterial Overgrowth (SIBO)

Small intestinal bacterial overgrowth (SIBO) is a condition that results from the colonization of the small intestine by increased numbers and/or abnormal types of microorganisms, typically bacteria that normally reside in the colon. The gut microorganisms (microbiota) of different segments of the GI tract are distinct, and they develop at and soon after birth increasing in quantity distally along the GI tract. Relatively low bacterial counts ensure that the small bowel does not face competition for nutrients following the ingestion of a meal. In the more densely populated colon, coliform bacteria, such as *Bacteroides*, are the primary energy source to maintain its integrity and function [6].

In healthy people, several defense mechanisms control the number and composition of small intestinal germs which normally make up the human intestinal microbiome and ecosystem. Typically, the stomach acid destroys many bacteria entering the gut with each meal or drink, while bile and pancreas juices and enzymes that are released in the intestine limit bacterial growth. Other key protecting mechanisms against small intestinal bacterial overgrowth are the migrating motor complex (which has a housekeeping function of gastrointestinal motility, pushing away and mixing the gut contents through a coordinated action of nerves and muscles in the entire gut reducing any luminal bacterial growth), the intestinal mucus that traps and fights bacteria, and the pylorus and ileocecal valves that inhibit any backward migration of bacteria. The intestinal microbiota affects the immune system and the absorption of nutrients, drugs, and toxic substances.

Depending on the disease stage, activity, duration, and the use of acid-blocking drugs, SIBO may be found in 30–50% of patients with PD [7]. In turn, such patients tend to experience more severe motor fluctuations (off-time, delayed on-time, and no on-time) than those without. SIBO induces an inflammatory state which might contribute to increased intestinal permeability that promotes the crossing of bacteria and their toxins into the bloodstream, triggering nerve ending activation and exacerbation of the neurodegenerative process in PD. SIBO may also contribute to slower stomach emptying and change the composition of the intestinal juice, the bacterial fermentation of sugars, and bile breakdown, further impairing levodopa absorption (Fig. 5.2).

Studies comparing the composition of the microbiome in the stools of patients with PD and matching healthy subjects have shown under- or overrepresentation of certain kinds of bacteria, some of which are associated with more severe motor symptoms, and others causing increased gut permeability and lower concentrations of ghrelin and vitamins, such as thiamine and folate.

A healthy small bowel maintains relatively low bacterial numbers because stomach acid, bile, pancreatic secretions, and other immunity factors block the growth of ingested and mouth bacteria, while intestinal peristalsis reduces stagnation. Disrupted motility in PD impairs the ability of the small bowel to clear bacteria. When microbial colonization occurs in the small bowel, it causes signs and symptoms of SIBO. Direct damage to the intestinal surface by bacteria may cause

Table: Gut microbiota and their implications in PD
Therapeutic manipulations of gut microbiota may improve levodopa pharmacokinetics
Gut microbiota influence levodopa absorption and may contribute to side effects
Dopaminergic medications may alter the gut microbiota composition
Gut microbiota strategies may have symptomatic and disease-modifying roles in PD

Fig. 5.2 Clinical implications of altered microbiome in Gut PD

malabsorption and loss of protein. Maldigestion of carbohydrates may lead to bloating, flatulence, and diarrhea from generated hydrogen and methane gases. Bacteria may also break down bile acids, causing fat malabsorption, weight loss, vitamin deficiencies, and osteoporosis.

Since patients with SIBO most commonly report nonspecific abdominal pain, distension, bloating, flatulence, and diarrhea, the diagnosis must rely on some objective evidence of abnormal numbers and types of bacteria in the small bowel. Although the gold standard for the diagnosis of SIBO is counting cultured bacteria from small intestinal aspirates, this method is invasive because it requires an endoscopy, and its diagnostic reliability is questionable unless it is performed in specialized centers. The lactulose breath test that measures breath hydrogen and methane after oral administration of lactulose is widely used. Some patients with SIBO harbor microbiota that produce methane gas rather than hydrogen; both hydrogen and methane need to be measured to increase the accuracy of diagnosis. Culturing aspirated small bowel fluid is invasive, risky, and costly and has limited sensitivity since SIBO is patchily distributed and may not be properly sampled. Importantly, the normal microbiome along the length of the gut remains to be completely and accurately defined. Breath tests for SIBO are based on the principle that the metabolism of starches, such as lactulose or glucose given by mouth, by bacteria that reside in the small bowel results in the production of hydrogen or methane that are then absorbed and exhaled as gas in an exhaled breath sample. Such testing is inexpensive, easily available, potentially administered at home, and noninvasive.

When given by mouth, lactulose is not normally absorbed in the small intestine and is, instead, metabolized by coliforms (bacteria normally residing only in the colon), to release hydrogen and methane gas. Patients with SIBO may thus have an early peak in exhaled hydrogen or methane due to lactulose metabolism by an abnormal small bowel flora. Conversely, glucose is normally absorbed in the proximal small bowel and is also metabolized by bacteria to hydrogen. Patients with SIBO may thus have elevated levels of exhaled hydrogen [8].

By convention, an increase of greater or equal to 20 parts per million (ppm) from baseline in hydrogen by 90 min is considered a positive test to diagnose SIBO. A level of greater than or equal to 10 ppm in methane is considered positive for methanogenic overgrowth. Although hydrogen is only produced by anaerobic bacteria normally found in the colon, methanogens are not actually bacteria but archaea, a species that do not cause disease. Thus, instead of SIBO, the term intestinal methanogen overgrowth (IMO) is used for a positive methane breath test, which has been associated with constipation because methane blocks intestinal transit. Ideally, in patients with PD who frequently suffer from constipation and bloating, both hydrogen (for SIBO) and methane (for IMO) should be measured during breath testing. Before the breath test is performed, antibiotics should be avoided for 4 weeks, and laxatives should be avoided for 1 week. Diet, smoking, and physical activity may also need to be altered [9] (Fig. 5.3).

Figure 5.3 shows a positive glucose breath test for hydrogen. There is a steady rise in parts per million of hydrogen in the exhaled breath during the first hour after

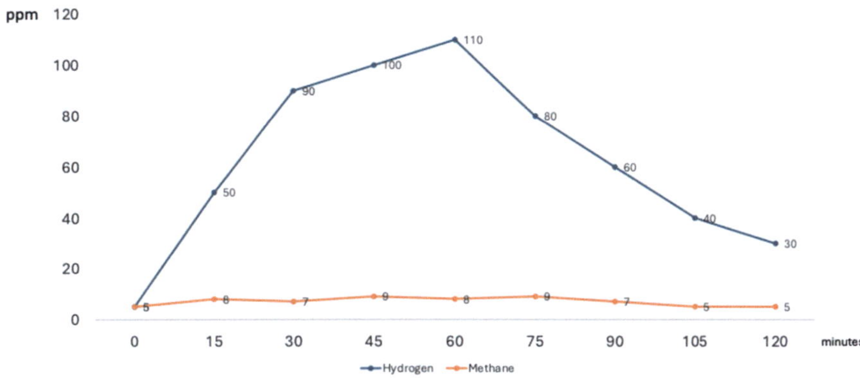

Fig. 5.3 Positive glucose breath test for hydrogen, but not methane

glucose ingestion, followed by a decline in the levels over the remaining time of the test. In contrast, the levels for methane remained flat throughout the entire period.

When SIBO or IMO is confirmed in patients with PD, an underlying cause (such as atrophic gastritis or PPI use) should be explored and corrected, if found, and consideration may be given to using drugs to stimulate intestinal motility to prevent recurrences. The multitude of microorganisms residing in the gut and their metabolic products can affect levodopa's absorption. Specifically, bacterial strains that are expressing tyrosine decarboxylase (TDC) may prematurely decarboxylate levodopa to dopamine in the small intestine before it gets absorbed, reducing the levels of bioavailable levodopa. Such patients may require increased levodopa dosing to achieve symptomatic control. The absorption of levodopa may also be affected by SIBO. Common predisposing conditions for SIBO in PD include underlying neurogenic dysmotility, longer disease duration, long-standing *H. pylori* infection, and concomitant proton pump inhibitor therapy, usually for relief of ongoing symptomatic gastroparesis. SIBO may lead to increased frequency of motor fluctuations, longer "off" phases, and suboptimal response to levodopa due to malabsorption. In turn, SIBO eradication with antibiotic treatment using rifaximin might improve motor performance in PD.

The most widely studied and used antibiotic for the treatment of SIBO is rifaximin 550 mg orally three times daily for 2 weeks, which acts in the gut lumen, is well tolerated, and improves motor fluctuations (off-time and delayed on-time). However, recurrences of SIBO are common in PD given the underlying abnormal gut motility contributing to stagnation of intestinal fluid and reaccumulating bacteria. Frequently, re-treatments for 2 weeks at a time every 1–3 months, or continuous, low dose (200 mg), are needed. Other antibiotics may be used for SIBO, given as 1–2 weeks every 1–3 months, as needed. However, rifaximin, which is nonabsorbable and works in the gut lumen, has likelihood of side effects and is generally preferable. At times, particularly for patients who are methane producers by breath testing, a combination of rifaximin and neomycin is suggested. Other antibiotics include ciprofloxacin (250 mg two times daily), norfloxacin (800 mg once daily), metronidazole

(250 mg three times daily), trimethoprim/sulfamethoxazole (160/800 mg two times daily), doxycycline (100 mg two times daily), amoxicillin–clavulanic acid (500 mg three times daily), and tetracycline (250 mg four times daily). For patients with IMO, neomycin is used (500 mg twice daily) for 2 weeks, alone or in combination with other antibiotics. Restoring healthy GI flora using a Mediterranean diet, low-starch diet, probiotics, prebiotics, and even fecal microbiota transplantation (FMT) may be considered in recurrent or resistant cases [10].

Recurrence rates for SIBO/IMO after treatment are high in patients with PD. Was the successful treatment based on symptom resolution alone or repeat breath testing with subsequent normalization? Were there any other, more objective signs of remission, such as weight gain, or normalization of nutritional deficiencies? To avoid the risks of intermittent antibiotic use, it is important to have a discussion of the limitations of breath testing and the risks of repeat or long-term antibiotic therapy.

Concomitant *H. pylori* and SIBO/IMO

Changes in the gut environment secondary to *H. pylori* infection or its treatment, such as reduction of stomach acidity, may alter the microbiota composition and facilitate proliferation of bacterial species in the small intestine leading to SIBO/IMO. Both *H. pylori* infection and SIBO have been associated with worsening in motor fluctuations in patients with advanced PD. The role of concomitant *H. pylori* infection and SIBO/IMO and their synergistic effect in PD is based on their partly overlapping mechanisms of action, and a synergistic effect is possible. *H. pylori* eradication therapy usually involves strong antibiotics used in combination that may partially or fully eradicate a concomitant undiagnosed SIBO/IMO. The opposite, that is, antibiotics for SIBO/IMO eradicating *H. pylori*, is unlikely.

Fecal Microbiota Transplantation for Constipation and SIBO in PD

Fecal microbiota transplantation (FMT) is a procedure that collects stool from a healthy donor and introduces it into the gastrointestinal tract of the patient with PD [11]. The procedure adds healthy bacteria into the recipient's intestines. FMT is usually performed by colonoscopy. The endoscopist guides the colonoscope to the proximal colon, and, as the instrument is withdrawn, the solution containing donor feces is dispersed into the lumen. Less commonly, FMT is delivered through a tube that is inserted through the nose into the small intestine. This approach does not require bowel preparation. FMT is regarded as safe and is well tolerated. The efficacy and safety of FMT for treating patients with PD has been recently explored and seems promising, since it may favorably improve both motor and non-motor symptoms of the disease, as measured by various well-validated scoring questionnaires. FMT is usually considered if patients have been found to have small intestinal overgrowth using breath hydrogen test and have not responded to antibiotics, or they have recurrences. In a recent placebo-controlled randomized trial, FMT through oral administration was clinically feasible and showed potential to improve the effectiveness of present medications in the clinical symptoms of PD patients [12]. Long-term outcomes and the need for repeat FMT have not been yet examined in PD.

References

1. Wei BR, Zhao YJ, Cheng YF, Huang C, Zhang F. *Helicobacter pylori* infection and Parkinson's disease: etiology, pathogenesis and levodopa bioavailability. Immun Ageing. 2024;21(1):1.
2. Lubomski M, Davis RL, Sue CM. Gastrointestinal dysfunction in Parkinson's disease. J Neurol. 2020;267(5):1377–88.
3. Menozzi E, Schapira AHV. The gut microbiota in Parkinson disease: interactions with drugs and potential for therapeutic applications. CNS Drugs. 2024;38(5):315–31.
4. Sun Q, Yuan C, Zhou S, Lu J, Zeng M, Cai X, Song H. *Helicobacter pylori* infection: a dynamic process from diagnosis to treatment. Front Cell Infect Microbiol. 2023;13:1257817.
5. Moss SF, Shah SC, Tan MC, El-Serag HB. Evolving concepts in *Helicobacter pylori* management. Gastroenterology. 2024;166(2):267–83.
6. Zafar H, Jimenez B, Schneider A. Small intestinal bacterial overgrowth: current update. Curr Opin Gastroenterol. 2023;39(6):522–8.
7. Varesi A, Campagnoli LIM, Fahmideh F, Pierella E, Romeo M, Ricevuti G, Nicoletta M, Chirumbolo S, Pascale A. The interplay between gut microbiota and Parkinson's disease: implications on diagnosis and treatment. Int J Mol Sci. 2022;23(20):12289.
8. Rao SSC, Bhagatwala J. Small intestinal bacterial overgrowth: clinical features and therapeutic management. Clin Transl Gastroenterol. 2019;10:e00078.
9. Mehravar S, Takakura W, Wang J, Pimentel M, Nasser J, Rezaie A. Symptom profile of patients with intestinal methanogen overgrowth: a systematic review and meta-analysis. Clin Gastroenterol Hepatol. 2024.
10. Rusch C, Beke M, Tucciarone L, Dixon K, Nieves C Jr, Mai V, et al. Effect of a Mediterranean diet intervention on gastrointestinal function in Parkinson's disease (the MEDI-PD study): study protocol for a randomised controlled trial. BMJ Open. 2021;11(9):e053336.
11. Sadowski K, Zając W, Milanowski Ł, Koziorowski D, Figura M. Exploring fecal microbiota transplantation for modulating inflammation in Parkinson's disease: a review of inflammatory markers and potential effects. Int J Mol Sci. 2024;25(14):7741.
12. Cheng Y, Tan G, Zhu Q, Wang C, Ruan G, Ying S, Qie J, Hu X, Xiao Z, Xu F, Chen L, Chen M, Pei Y, Zhang H, Tian Y, Chen D, Liu X, Huang H, Wei Y. Efficacy of fecal microbiota transplantation in patients with Parkinson's disease: clinical trial results from a randomized, placebo-controlled design. Gut Microbes. 2023;15(2):2284247.

Drug Therapies for Non-motor Symptoms and Gut PD

6

Non-motor symptoms significantly contribute to the burden of PD, but evidence for effective treatment is very limited. Hence, off-label medications are commonly used, and comprehensive multidisciplinary care is essential. Many non-motor symptoms worsen with advancing disease or with dopaminergic therapy. The development of PD-related dementia may not only decrease the response to treatment but also affect treatment adherence. Depression and anxiety can be treated with selective serotonin reuptake inhibitors, selective serotonin–norepinephrine reuptake inhibitors, or less commonly dopamine agonists, with careful attention to drug interactions. Prior to decision-making on drug therapies for gut PD, the possible effects of antiparkinsonian drugs on gut function, the overall burden of disease, and patient-related factors need to be considered. Depression and anxiety can be treated with selective serotonin reuptake inhibitors, selective serotonin–norepinephrine reuptake inhibitors, or less commonly dopamine agonists, with careful attention to drug interactions and the potential development of the serotonin syndrome. Psychiatric symptoms of hallucinations and delusions can be treated with atypical antipsychotic medications (clozapine or quetiapine). Other dopamine D2 receptor-blocking antipsychotic agents are not used because they can worsen Parkinsonism.

Effects of Antiparkinsonian Drugs on Gut Function

Levodopa (L-dopa), the main precursor in dopamine synthesis, is the most effective drug for the treatment of PD, since it improves motor function and quality of life. Unfortunately, it needs frequent dosing and has the highest risk of dopamine side effects, such as "wearing off" and dyskinesia. Although in early PD levodopa is best taken with meals to minimize nausea, later, it is best taken on an empty stomach, since protein in the food may block its effect.

Levodopa is often combined with carbidopa, a peripheral decarboxylase inhibitor that blocks its conversion to dopamine in the bloodstream and prevents nausea, vomiting, and low blood pressure upon standing. By preventing such conversion to

G. Triadafilopoulos, *The Gut in Parkinson's Disease*,
https://doi.org/10.1007/978-3-031-77171-2_6

dopamine, which does not cross into the brain, carbidopa allows a smaller amount of levodopa to be given and achieve its effect in the brain.

Treatment is best started with an immediate release formulation of carbidopa–levodopa that dissolves on the tongue and can be taken without water, aiming for the lowest clinically effective dose. Controlled-release (CR) tablet levodopa preparations are available but not recommended as initial therapy, but they can be used later for the convenience of twice-daily dosing since they reach the brain more slowly over time. CR tablets are not well absorbed and typically require a 30% increase in dose to achieve a similar clinical effect. CR tablets are sometimes better tolerated in patients with nausea.

Levodopa is ubiquitously used, either alone (monotherapy) or together with other drugs. Unfortunately, it needs frequent dosing and has the highest risk of dopamine side effects, such as "wearing off" and dyskinesia. Although in early PD levodopa is best taken with meals to minimize nausea, later, it is best taken on an empty stomach, since protein in the food may block its effect. Levodopa is often combined with carbidopa, a peripheral decarboxylase inhibitor that blocks its conversion to dopamine in the bloodstream and prevents nausea, vomiting, and low blood pressure upon standing. By preventing such conversion to dopamine, which does not cross into the brain, carbidopa allows a smaller amount of levodopa to be given and achieve its effect in the brain (Fig. 6.1).

Other than levodopa, which remains the cornerstone of therapy, there are many drugs that are used in the management of PD, also aiming at improving neurological function. The mechanisms of action, their magnitude, and duration of their effects vary. Dosing and tolerability also vary among patients. Unfortunately, many of them have gut adverse effects, particularly constipation. Often these drugs are used in

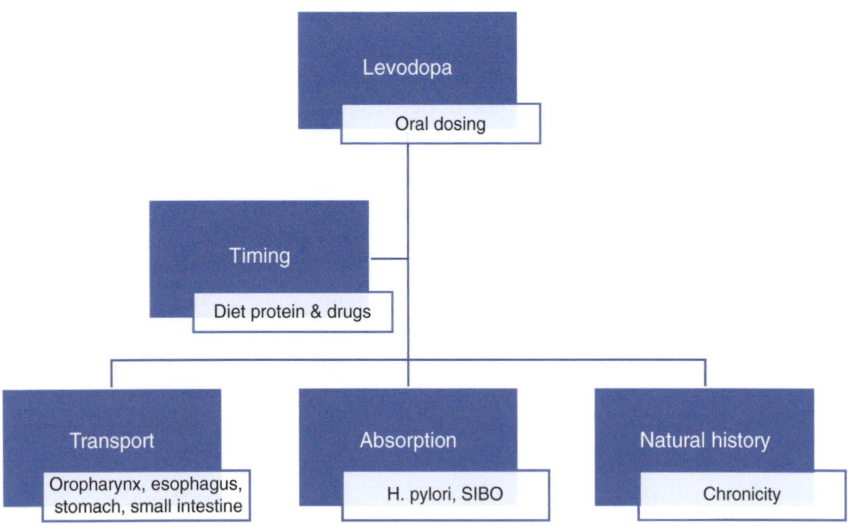

Fig. 6.1 Outline of key gut alterations affecting levodopa pharmacodynamics

Drug	Gut side effect(s)	Mechanism of action	Frequency (%)
Levodopa/ Carbidopa	Constipation, nausea	Precursor of dopamine synthesis/ peripheral decarboxylase inhibitor	22-30
Pramipexole	Constipation, nausea	Dopamine agonist	14-28
Ropinirol	Nausea	Dopamine agonist	60
Rotigotine	Nausea	Dopamine agonist	15-48
Amantadine	Constipation, dry mouth	Dopamine agonist	13-16
Rasagiline	Indigestion	MAO Type B Inhibitor	7
Entacapone	Diarrhea	COMT inhibitor	10
Trihexyphenidyl	Nausea, dry mouth	Anticholinergic	30-50
Selegiline	Nausea	MAO type B inhibitor	10
Tolcapone	Nausea	COMT inhibitor	30-50
Rivastigmine	Nausea	Acetylcholinesterase inhibitor	23-47

Fig. 6.2 Gastrointestinal side effects of various anti-Parkinson's drugs and their mechanisms of action

combination with levodopa, or other drugs, raising the possibility of drug–drug interactions. Upon initiation of a new drug therapy, as well as periodically, all patients should try to keep a diary of their gut symptoms on a weekly basis and share this information with their movement disorder specialist who, in turn, may need to change the drug, their dose, or the frequency of use, based on careful and ever-changing benefit-to-risk assessment. Trade-offs are frequently needed and negotiated among the treating physicians. Further, alternative, nondrug approaches may need to be considered.

Figure 6.2 shows the most common drugs for PD, their gut-related side effects, their mechanisms of action, and their relative frequency of occurrence.

Duopa Therapy

As PD advances, patients may notice that their levodopa pills are not working as well as they used to, and they may need to take higher doses or take their pills more often. Dyskinesia occurs when the body has too much levodopa and patients get twisting or turning movements. The diagram below outlines the levodopa absorption and pharmacodynamics as they relate to clinical manifestations. "Off" time is the period when levodopa is not working well, and patients feel stiff, move slowly, or cannot get around as well. "On time" is when the levodopa is working well to control PD symptoms. Such fluctuations of "off" and "on" times are common (left), particularly in advanced PD. With time, it gets harder to find the "sweet spot" where the levels are at a steady state (right), controlling symptoms. Duopa is delivered into the intestine, so levodopa can be absorbed efficiently. To initiate Duopa therapy, an endoscopist makes a small hole (called a "stoma") in the stomach wall and places a medication delivery port with its tip in the small intestine for direct delivery through a portable pump and absorption, bypassing the stomach. Duopa therapy delivers levodopa continuously over 16 h. For whatever reasons, such as intolerance or lack of efficacy, the medication delivery port can be completely removed endoscopically so the stoma heals. Typically, patients take a Duopa cassette out of the refrigerator and let it sit for 20 min at room temperature. They then connect the pump to the cassette and then to the stomach tube and then start the pump to get a morning dose.

Continuous dosing starts automatically for 16 h. After 16 h, the patient disconnects the pump, flushes the stomach tube with drinking water, and discards the Duopa cassette. They can then take a nighttime dose of oral carbidopa/levodopa pills as prescribed. The movement disorder specialist may also decide if patients need extra medicine during the day. If the patient feels like they are "wearing off," they can give themselves an extra dose through the pump. For ambulation, Duopa carrying cases come in many different styles, such as fitted belt, hip pack, lightweight vest, and fitted vest, and patients may choose the style that works best for their needs. In a multicenter study on long-term effects of such therapy, improvement of motor and non-motor fluctuations and some non-motor symptoms, an increase of dyskinesias with a decrease of the troublesome dyskinesias was associated with a high rate of dropout for lack of efficacy or adverse events related to the drug [1] (Figs. 6.3, 6.4).

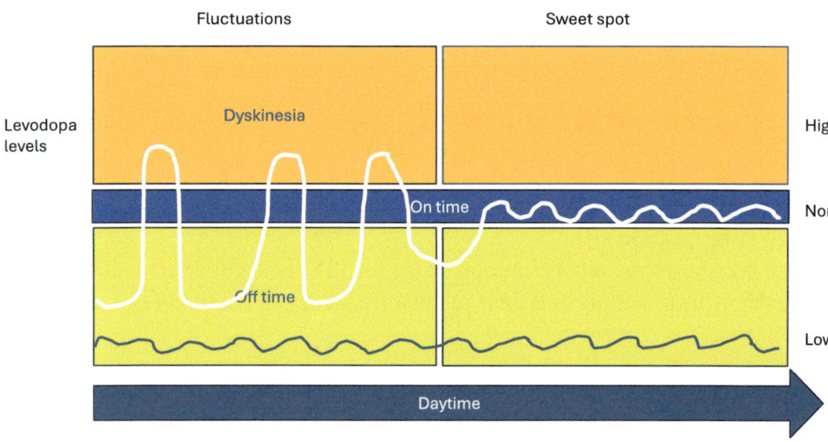

Fig. 6.3 Levodopa absorption and pharmacodynamics as they relate to clinical manifestations of PD

Fig. 6.4 Endoscopic image of the stoma through which the Duopa catheter is introduced to ultimately deliver steady levels of drug for small intestinal absorption

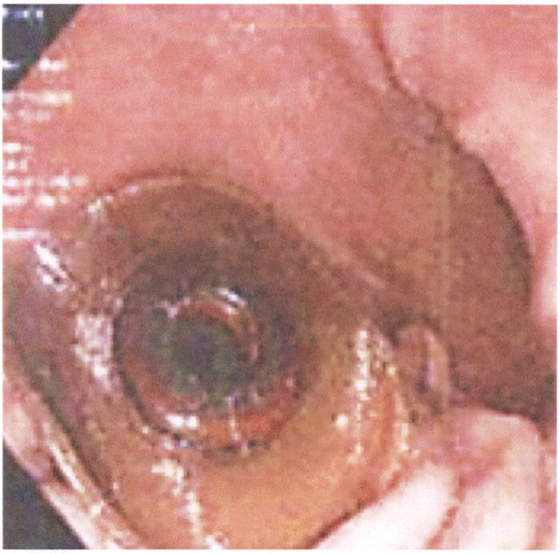

Endoscopic image of a gastrostomy that allows for passing the Duopa catheter (not shown) into the intestine for direct delivery of levodopa, bypassing the stomach. A gastrostomy is created endoscopically and provides a continuous communication between the skin of the anterior abdominal wall and the body of the stomach. This connection is made to allow the placement of the Duopa catheter under endoscopic guidance into the intestine. The patient may continue using oral feedings, as needed.

Pharmacotherapy of Gut PD

Table highlights the various agents used in the pharmacologic management of gut PD. Although these agents are used extensively in PD patients, they have not been formally tested or approved for safety, tolerability, and efficacy, and their impact on PD-related outcomes is uncertain (Fig. 6.5).

Analgesics

Abdominal pain may be very difficult to manage because many pain medications slow intestinal transit. Management by a physician specialized in pain management is often necessary. Analgesics typically prescribed for neuropathy, such as tricyclic antidepressants (amitriptyline, nortriptyline, and imipramine), and antiepileptic medications such as gabapentin and pregabalin can be used with caution. Opioids should be avoided because their side effects (e.g., sphincter of Oddi dysfunction, constipation) may worsen symptoms. Transdermal buprenorphine has a high receptor affinity combined with a high potency and a low risk for euphorigenic effects that may be used in selected patients.

Coating Agents/Sucralfate

Sucralfate, 1 g orally qid, alone or in combination with a proton pump inhibitor is used in the treatment of drug-induced esophageal injury, a frequent occurrence in patients with PD [2]. Sucralfate binds to the injured tissue, thereby delivering growth factors and reducing access to pepsin and acid, thereby promoting healing. However, sucralfate may significantly interfere with the pharmacokinetics of levodopa, and careful monitoring is needed.

Prokinetics

Prokinetics are drugs which enhance gastrointestinal motility by increasing the frequency or strength of gastrointestinal contractions without disrupting their rhythm. These drugs are frequently used in the management of PD patients with dysphagia, heartburn, abdominal bloating, nausea, vomiting, early satiety, and constipation.

Table	Agents and dosage ranges
Analgesics	Acetaminophen Non-steroidal anti-inflammatory drugs (NSAIDs) Gabapentin/pregabalin Opioids
Coating agents	Sucralfate
Prokinetics	Dopamine receptor antagonists (Metoclopramide, Domperidone) Motilin agonists (Erythromycin, Clarithromycin) Ghrelin receptor agonists (Relamorelin) 5-HT4 receptor agonists (Prucalopride)
Antiemetics	Phenothiazines (Promethazine, Prochlorperazine) 5-HT3 receptor antagonists (Ondasetron, Granisetron) Neurokinin-1 receptor antagonists (Aprepitan)
Laxatives	Osmotics (PEG, Lactulose) Stimulants (Senna) Magnesium Chloride channel activators (Lubiprostone) Guanylate cyclase receptor activators (Linaclotide, Plecanatide) Sodium/hydrogen exchange-3 inhibitors (Tenapanor) Misoprostol Peripherally acting mu receptor antagonists (PAMORA) (Methylnatrexone, Naloxegol, Alvimopan)
Antidiarrheals	Loperamide Diphenoxylate with atropine Tincture of opium Octreotide 5HT-3 antagonists (Alosetron) Mixed opiod receptor antagonists (Eluxadoline)
Antispasmodics	Dicyclomine Hyoscyamine Peppermint oil
Antibiotics	Rifaximin Neomycin
Acid blockers	Antacids H2-receptor antagonists Proton pump inhibitors
Bile acid sequestrants	Cholestyramine, Colestipol
Topical anti-inflammatory	Budesonide
Neuromodulators	TCA (Tricyclic antidepressants) SSRI (selective serotonin reuptake inhibitors) SNRI (serotonin and noradrenaline reuptake inhibitors) TNAS (tetracyclics or noradrenergic and specific serotonergic antidepressants) AAP (Atypical antipsychotics)
Mineralocorticoids	Fludrocortisone
Vasopressors	Midodrine
Acetylcholinesterase inhibitors	Pyridostigmine

Fig. 6.5 Various agents used in the pharmacologic management of Gut PD

Examples of these agents include the dopamine receptor antagonists, such as metoclopramide and domperidone; motilin agonists, such as erythromycin and clarithromycin; ghrelin receptor agonists, such as relamorelin; or 5-HT4 receptor agonists, such as prucalopride. Unfortunately, there are significant drug–drug interactions that limit the utility of these agents in patients with gut PD, and careful assessment and monitoring is needed before these agents can be used [3].

Antiemetics

These drugs are used mostly for the control of nausea and vomiting, either acutely or chronically. Several classes of antiemetics may antagonize the neurotransmitter receptors involved in the induction of nausea and vomiting, and they are classified according to their primary action [M1, muscarinic; D2, dopamine; H1, histamine; 5-hydroxytryptamine (HT)-3, serotonin; and neurokinin-1 (NK1) receptor, substance P]; some agents affect multiple receptors. Drug–drug interactions limit the utility of these agents in patients with gut PD, and careful assessment and monitoring is needed before these agents can be used.

Laxatives

Many substances that stimulate bowel movements, so-called laxatives, are available over the counter, and they are safe and inexpensive without evidence that any one is better than any other. The treatment of constipation should be individualized, depending on patient preferences or comorbidities and considering the presence or not of fecal impaction. Glycerol or bisacodyl suppositories may be effective in patients with stool in the rectum but will not affect the whole colon. Rectal interventions should be reserved as a last resort since they are burdensome for the patient and caretakers. Polyethylene glycol 3550-PEG, also available over the counter, promotes a dose-dependent bowel cleansing by pulling water into the gut lumen and promoting soft stools without increasing gut motility. To be effective, an optimal dose needs to be found and used daily, not as needed. Such dose can be titrated with fecal incontinence being the dose-limiting factor. If that is reached, slow reduction in the daily dose usually suffices.

PEG is minimally absorbed in the gastrointestinal tract and causes water retention in the stool which softens the stool, and bowel movements occur more frequently [4, 5]. Overall, polyethylene glycol is preferred over lactulose [6].

Poorly absorbed or nonabsorbable sugars and saline laxatives cause intestinal water secretion and therefore increase stool frequency. Excessive use of these agents may result in electrolyte and volume overload in patients with renal and cardiac dysfunction [7]. Magnesium and sulfate both individually exert a laxative action, mainly mediated by an osmotic effect due to their incomplete absorption in the gastrointestinal tract [8]. Daily consumption of 1 L magnesium sulfate-rich natural mineral water reduced constipation and hard or lumpy stools in women with constipation [9].

Lubiprostone, 8–24 mcg bid after meals, is a locally acting chloride channel activator that enhances chloride-rich intestinal fluid secretion and intestinal motility, resulting in a laxative effect [10]. Lubiprostone may lead to persistent improvement of abdominal bloating, discomfort, stool frequency, straining, and constipation for 6–12 months. The most common side effect is dose-dependent nausea, occurring in up to 30% of patients. Because of this, it may not be preferred in patients with gut PD and gastroparesis, but it carries no drug–drug interactions with levodopa. The approved dose is 24 mcg taken twice daily with food, which decreases the frequency of nausea, although 8 mcg and 12 mcg can be tried [11].

Linaclotide, 72–290 mcg orally daily, is a polypeptide that binds to guanylate cyclase C receptors located on the luminal surface of the gastrointestinal tract, causing a cyclic-GMP-mediated increase in chloride and bicarbonate secretion while inhibiting absorption of sodium ions, therefore increasing the net secretion of fluid in the gut lumen. Dose titration and optimization to every other day use may be tried. Although there are no drug interactions with levodopa, there are no formal trials of this drug in PD [12, 13]. Compared to placebo, oral linaclotide 145 mcg and 290 mcg daily leads to significantly higher rates of >3 complete spontaneous bowel movements (CSBM) per week in patients with chronic idiopathic constipation [14]. In some PD patients who develop diarrhea and cramps, 72 mcg daily may be effectively used [15].

Plecanatide, 3 mg orally daily, is also a minimally absorbed peptide agonist of the guanylate cyclase C receptor that stimulates intestinal fluid secretion and transit and is approved for the treatment of chronic constipation but not specifically studied in patients with gut PD. It has no drug interaction potential with levodopa [16]. Patients treated with plecanatide (3 mg or 6 mg once daily) experience improvement in the frequency of complete spontaneous bowel movements with rare diarrhea as side effect [17, 18].

Prucalopride, 1–4 mg orally daily, is a 5HT4 prokinetic agent that is safe and well tolerated even in older patients with PD, causing improvement of slow transit constipation and quality-of-life scores for up to 24 months. It stimulates contractions in colonic longitudinal smooth muscles by promoting acetylcholine release via activation of 5-HT_4 receptors on presynaptic, cholinergic enteric neurons. They all increase spontaneous bowel movements and improve bloating, discomfort, stool frequency, and straining. The drug has not been formally studied in PD but carries no potential interactions with levodopa [19–21]. Abdominal pain or bloating, diarrhea, dizziness, gas, and fatigue are possible side effects. When cost is an issue, PEG 3350-electrolye solution, with its lower cost, may be tried, and it has been shown as effective [22].

Tenapanor, 20–50 mg daily, is a potent small-molecule inhibitor of the sodium/hydrogen exchanger isoform 3 (NHE3) localized in the apical membrane of intestinal epithelial cells, which is minimally absorbed and acts locally to inhibit sodium absorption. It also modulates tight junctions in the small intestine and increases transepithelial electrical resistance (TEER) reducing phosphate ion permeability and paracellular phosphate absorption in the presence of NHE3. The resulting increase of sodium in the gut leads to increased fluid secretion [23].

Misoprostol, 100–200 mcg four times daily (or less), is a prostaglandin analog at times used for NSAID-induced gastric ulcer prevention and may also treat some patients with severe constipation [24]. Increasing doses or frequency may increase unnecessarily abdominal cramping.

Colchicine, 0.3–0.6 mg once daily, a drug used for gout, may be effective in titrated doses for the treatment of chronic constipation in refractory patients with gut PD [25]. Colchicine should not be used in patients with impaired kidney function.

The peripherally acting mu-opioid receptor antagonists (PAMORA) are derivatives of naloxone or naltrexone (methylnaltrexone, naloxegol, and naldemedine) that do not cross the blood–brain barrier, and they are a new treatment option for patient with PD who may suffer from opioid-induced constipation (OIC). Importantly, PAMORA do not reverse the analgesic effect of opioids, but by preferentially binding to opioid receptors in the gut and significantly increasing the frequency of spontaneous bowel movement, they effectively relieve constipation and improve quality of life [26]. PAMORA affect all peripheral tissues which may help the management of gastroparesis, opioid-induced dysphagia, and opioid-induced xerostomia [27]. The American Gastroenterological Association strongly recommended traditional laxatives as first-line agents and PAMORA as superior to no treatment, particularly using naldemedine and naloxegol and, less so, methylnaltrexone [28, 29]. Methylnaltrexone does not permeate the BBB and is poorly absorbed and is given subcutaneously with weight-based dosing of 8 mg or 12 mg every other day and is effective and well-tolerated treatment with 50–60% of patients experiencing laxation within 4–24 h after dosing, twice as much than placebo [30].

Naloxegol is an orally administered naloxone derivative, with proven effectiveness in OIC [31]. Adverse effects are infrequent (<10%), primarily mild to moderate, mainly abdominal pain, diarrhea, flatulence, and nausea.

Naldemedine is a potent antagonist of mu, delta, and kappa opioid receptors that is not crossing the brain barrier. It is administered in 0.2 mg tablets once daily, regardless of meals. In a meta-analysis of six randomized control trials with 2991 patients, the odds ratio (OR) for spontaneous bowel movement response was 3 in favor of naldemedine [32]. Naldemedine does not affect opioid-mediated analgesia and is not associated with symptoms of opioid withdrawal syndrome. Axelopran is a highly potent peripheral mu-opioid receptor antagonist with limited ability to cross the brain barrier. In non-cancer pain, axelopran appears better than placebo with a durable increase in bowel movement frequency and return toward normal bowel function [33].

The classical mu-opioid receptor antagonists, naloxone and naltrexone, revert all the peripheral and central opioid effects, including analgesia, since they cross the brain barrier. Naloxone has antagonistic effect on opioid receptors in the gut, with minimal impact on analgesia. There are several oral combinations of naloxone with oxycodone and pentazocine for pain treatment. There is also a naloxone/buprenorphine sublingual tablet to treat opioid addiction and a hydromorphone/naloxone, prolonged-release tablet formulation. Slow-release tablets, containing 5, 10, and 20 mg of naloxone administered once daily for OIC in non-cancer patients show significant increase in laxation twice as much as placebo [34]. Naltrexone is available in combination with morphine or oxycodone tablets for severe pain treatment.

Oxycodone/naloxone in fixed-ratio 2:1 dose extended-release tablets is widely used to treat severe pain, with the premise that naloxone counteracts OIC by blocking the action of oxycodone at opioid receptors locally in the gut.

Alvimopan is indicated only for the treatment of postabdominal surgery ileus and is available only to hospitals for short-term inpatient use because of an increased risk of ischemic cardiovascular events associated with long-term therapy. It is administered in 12 mg capsules once prior to surgery and up to 7 days postoperatively in hospitalized patients [27].

Drugs for Fecal Incontinence

Many patients with gut PD and many elderly are unable to voluntarily initiate voiding; many require bladder catheters to void urine and "manual bowel programs" with digital rectal stimulation and manual extraction to void stool. If unsuccessful, fecal incontinence occurs. These manual bowel programs are time-consuming (1–2 h) and stigmatizing and cause rectal pain and discomfort. Laxatives and enemas produce defecation, but onset and duration are unpredictable, prolonged, and difficult to control, which can produce involuntary defecation and fecal incontinence. Safe, effective, on-demand, rapid-onset, short-duration, drug-induced, voiding therapy could eliminate or reduce need for manual bowel programs and colostomies in patients who are unable to voluntarily initiate evacuation [35].

Antidiarrheal Agents

In general, the use of antidiarrheal agents is best avoided in patients with unclear etiology of diarrhea, particularly if infectious or inflammatory causes have not been fully evaluated. Loperamide is a synthetic opioid that reduces peristalsis by inhibiting the release of acetylcholine through activation of the mu-opioid receptors, which slows GI transit time, allowing more time for absorption of water in the intestine in turn leading to a reduction in stool frequency [36, 37].

Inhibition of acetylcholine release also leads to antisecretory activity because muscarinic acetylcholine receptors exist on secretory intestinal epithelial cells [38]. Loperamide treatment delays both small bowel and whole gut transit. Loperamide also has a relaxing effect on localized and segmental colonic contractions improving stool consistency and reducing pain intensity [39].

Diphenoxylate is a synthetic opioid that, by acting on the presynaptic opioid receptors (predominantly mu receptors) in the enteric nervous system, it inhibits motility and decreases segmental contractions, thereby prolonging gastrointestinal transit time. It also reduces the epithelial secretion of fluid and electrolytes and enhances active absorption by mild action on delta receptors. Commercial preparations contain a subtherapeutic amount of atropine to discourage abuse. High doses of diphenoxylate–atropine can result in symptoms of excess cholinergic and central narcotic effects, such as drowsiness, flushing, dry mouth, tachycardia, dilated pupils, rash, and nausea [40]. In certain patients, idiopathic bile acid malabsorption may be the cause of diarrhea, and treatment with a bile acid binder cholestyramine may be effective [41].

5-HT3 Antagonists

Alosetron slows down fecal transit through the colon and provides more time for water absorption, resulting in a decrease in total moisture and volume of fecal contents. Contraindications to the drug include history of chronic or severe constipation, strictures of the GI tract, history of perforation or adhesions, ischemic colitis, inflammatory bowel disease, diverticulitis, and severe hepatic impairment.

Mixed Opioid Receptor Antagonists

Eluxadoline is a μ- and κ-opioid receptor agonist and δ-opioid receptor antagonist that acts locally on the gut mucosa and enteric nervous system. Local μ-receptor agonism reduces colonic secretion and slows gut transit, while δ-opioid receptor antagonism counteracts any excessive μ-receptor inhibition and prevents excessive constipation from unopposed μ-receptor stimulation. In addition, concomitant μ-agonism and δ-antagonism have synergistic analgesic effects, relieving associated abdominal discomfort in patients with IBS. Kappa-opioid receptor agonism reduces visceral hypersensitivity [42].

Octreotide, is a synthetic somatostatin analog that, when given intravenously, subcutaneously, or as a depot injection, diminishes diarrhea by acting directly on epithelial cells to reduce the secretion of pancreatic and GI hormones, such as VIP, serotonin, gastrin, secretin, and pancreatic polypeptide. Octreotide is particularly useful in patients with neuroendocrine tumors since it prolongs intestinal transit time, promotes intestinal absorption, and decreases secretion of fluids and electrolytes [43, 44].

Antispasmodics

Dicyclomine, 10–20 mg orally four times daily, is effective in controlling abdominal pain, cramping, and urgency, and it may be used as needed but has significant anticholinergic properties that may limit its use because of side effects, such as blurred vision, dizziness, dry mouth, and impaired cognitive function, particularly in older adults.

Hyoscyamine, 0.125–0.25 mg sublingually or orally every 4–8 h or as needed, maximum 1.5 mg/day or extended release 0.375–0.75 mg orally every 12 h or 0.375 mg every 8 h, is effective in controlling abdominal pain, cramping, and urgency, and it may be used as needed but also has significant anticholinergic properties that may limit its use because of blurred vision, dizziness, dry mouth, and impaired cognitive function, particularly in older adults.

Peppermint oil is a widely used therapy for crampy abdominal pain and bloating. Although predominantly an antispasmodic, it has antinociceptive, anti-inflammatory, and carminative effects [45].

Antibiotics

Small intestinal bacterial overgrowth (SIBO) is almost universally present in CIPO, causing abdominal distension and promoting intestinal bacterial translocation. Although not approved for this indication, rifaximin, a nonabsorbable antibiotic, can be used to treat microbial overgrowth. Alternative or sequential treatments with other antibiotics may be offered. For SIBO due to methane-producing bacteria, oral neomycin may be preferred [46].

Neuromodulators

Neuromodulators, widely tested as effective in patients with irritable bowel syndrome, work by enhancing the synaptic transmission of 5-hydroxytryptamine, noradrenalin, and dopamine, leading to desensitization of their postsynaptic receptors along the brain–gut axis. These agents, in doses lower than those used in psychiatry, not only are useful in treating psychiatric comorbidities but importantly modifying gut motility, improving central downregulation of visceral signals in irritable bowel syndrome (IBS) [47]. Typically, the first-line treatment for pain management in IBS is using tricyclic antidepressants or the serotonin and noradrenaline reuptake inhibitors. When selecting a neuromodulator to treat IBS, clarifying the predominant bowel symptom is important; selective serotonin reuptake inhibitors help constipation, not pain, but may cause diarrhea, while tricyclic antidepressants help diarrhea but may cause constipation. A clinical response may occur in 6–8 weeks, but long-term treatment is often needed to prevent relapse. Augmentation therapy may also be needed. Table shows the various options and dose ranges for patients with IBS (Fig. 6.6).

Although constipation is the most common gastrointestinal symptom in patients with Parkinson's disease, trials to assess efficacy of various drugs have been scarce. A non-blinded study in ten patients with PD showed that an osmotic laxative (polyeth 3350) was efficacious. A 2-week randomized, placebo-controlled trial in 54 patients showed that lubiprostone, a secretagogue and chloride channel activator, is efficacious without substantial adverse effects, although nausea is common.

Prucalopride and linaclotide, a guanylate cyclase C agonist luminally active and minimally absorbed gastrointestinal secretagogue, have not been studied in patients with Parkinson's disease. In view of the few trials in patients with PD with

Table	Neuromodulating agents and dosage ranges
TCA (Tricyclic antidepressants)	Amitriptyline 25-150 mg qd Imipramine 250150 mg qd Desipramine 25-150 mg qd Nortriptyline 25-150 mg qd
SSRI (selective serotonin reuptake inhibitors)	Citalopram 10-40 mg qd Escitalopram 5-20 mg qd Fluoxetine 10-40 mg qd Paroxetine 10-40 mg qd Sertraline 50-150mg qd Fluvoxamine 50-150mg qd
SNRI (serotonin and noradrenaline reuptake inhibitors)	Venlafaxin 750225 mg qd Duloxetine 30-90 mg qd Milnacipran 50-100 mg bid
TNAS (tetracyclics or noradrenergic and specific serotonergic antidepressants)	Mirtazapine 7.5-45mg qhs
AAP (Atypical antipsychotics)	Quetiapine 25-200 mg qd Olanzapine 2.5-10mg qd Brexpiprazole 1-1.5 mg qd Aripiprazole 2.5-5 mg qd

Fig. 6.6 Caption. Modified from: Hanna-Jairala I, Drossman DA. Central Neuromodulators in Irritable Bowel Syndrome: Why, How, and When. Am J Gastroenterol. 2024 Jul 1;119(7):1272–1284. https://doi.org/10.14309/ajg.0000000000002800. Epub 2024 Apr 9. PMID: 38595149; PMCID: PMC11208063.

constipation, we recommend the use of the same treatment algorithm as in patients with idiopathic chronic constipation. In patients with dyssynergic defecation, although botulinum neurotoxin injections into the anal sphincter have been recommended, further study is needed before recommending this approach as standard practice. Safe and easy access to the toilet, with pertinent space improvements, based on ADA recommendations, is pivotal to avoid accidents and falls. The benefits of acupuncture and the effectiveness and safety of adjunctive traditional Chinese herbs for constipation in PD are questionable.

Diet, Lifestyle, and Non-pharmacologic Intervention for Constipation

Treatment of Constipation in PD

Prior to any treatment decisions regarding the patients' constipation, an effort should be made—together with the movement disorder specialist—to optimize and stabilize the overall neurological picture utilizing the most effective and practical drug regimen. This way, any adverse effects of anti-Parkinson's drugs, with their impact on anorectal function, will be considered. The treatment should then be individualized, considering the patient's frequency and severity of abdominal pain, bloating, straining, incomplete evacuation, and urge incontinence and the underlying pathophysiology of constipation based on functional testing, age, comorbid conditions, and patients' concerns and expectations. It is not uncommon to encounter patients treating constipation as needed, not using a regular regimen that works preemptively, and not as a reaction to the need for evacuation. The treatment for dyssynergic defecation consists of (1) standard treatment for constipation, (2) single or multipronged pharmacologic therapy, (3) vibrating capsule therapy, (4) biofeedback therapy tailored to previously identified anorectal manometry ROM and MR defecography findings, (5) treatment of IMO, and (6) other measures, such as anal and puborectalis muscle botulinum toxin injection, rectocele repair, posterior rectopexy, end colostomy, or ileostomy.

Baseline Treatment

This involves a detailed clinical assessment and correction of coexisting issues, such as avoiding constipating medications, adequate fiber (up to 25 g/day) and fluid intake, and regular exercise. Overreliance on fiber is common and frequently causes abdominal distention and pain in patients with PD. Some lifestyle modifications can be useful [48]. In addition, patients should be encouraged to capitalize on mechanisms that stimulate the colon [49], such as timing after waking and after a meal and avoiding postponing defecation, as the urge subsides after a few minutes and may not return for several hours. Patients should receive instructions regarding timed toilet training and effective straining methods. Programmed defecation and timed toilet training consists of educating the patient to attempt a bowel movement at least twice a day, usually 30 min after meals. During attempted defecation, they must be instructed to push at a submaximal effort of straining and for no more than 5 min. It is important to emphasize that stool impaction should be prevented using regular digital disimpaction.

Fiber Supplements

Fiber intake accelerates colonic transit and bulks the stool either by drawing fluid into the colon or by facilitating fermentation and favorably altering gut microbiota or epithelial permeability [50]. Fiber is useful in patients with mild to moderate constipation associated with PD—particularly those with slow transit—because it decreases straining, increases stool frequency, softens stool consistency, minimizes fecal incontinence, and decreases the need for laxatives. If taken with adequate amounts of water, soluble fiber, such as psyllium; insoluble fiber, such as calcium polycarbophil, bran and methylcellulose, or mixed fiber supplements; and natural dietary fiber, such as prunes, are effective. Patients should not expect any improvement for several weeks of regular fiber intake, and they should be prepared to reduce the dosage if bloating, flatulence, or abdominal discomfort occurs [51]. Patients with fecal impaction or those confined to the bed or those on fluid restricted diet should not be given fiber supplements [52].

Pharmacologic Approaches

Although laxatives, intestinal secretagogues and serotonergic enterokinetic agents, are used in most patients with PD and constipation, they have not been fully evaluated in patients with PD who mostly suffer from dyssynergic defecation. Nevertheless, these drug therapies can be used as adjuncts along with biofeedback therapy, especially in patients with slow colonic transit.

Laxatives

Any substance that stimulates bowel movements can be called a "laxative." However, new classes of medications, such as intestinal secretagogues and selective 5-HT agonists, are usually distinguished from the traditional agents that are generally accessible over the counter, safe, and inexpensive. There is no evidence for any one laxative being superior to any other. Unfortunately, regular laxatives may not be effective, since the colon in PD is like a rigid pipe that is unresponsive to stretching and with a constricted outlet. Fiber-rich bulk supplements typically cause further bowel impaction and are not recommended. Osmotic laxatives, such as polyethylene glycol 3550-PEG, promote a dose-dependent bowel cleansing effect by pulling water into the gut lumen and ensuring appropriately soft stool consistency are usually the first-line therapy, but they do not increase gut motility, nor release sphincter malfunction. Stimulant laxatives by mouth are commonly used, but over time increasing doses are needed. Glycerol or bisacodyl suppositories may be effective in patients with stool in the rectum but will not affect the rest of the large bowel. Rectal or trans-anal irrigation using enemas may benefit patients with stool in the distal colon, above the rectum. Unfortunately, rectal interventions are highly burdensome for patients and caretakers, since they are uncomfortable and messy, and they are usually reserved as a last resort for those patients who have not responded to other interventions. In severe cases, daily, programmed evacuation, combining suppositories, enemas, and manual stimulation and stool disimpaction are appropriate.

Other pharmacological options using linaclotide, lubiprostone, and prucalopride for constipation, by adding or switching them to ongoing laxative therapy if patients do not respond to first-line conventional laxatives.

Linaclotide, available as 145 and 290 μg tablets for chronic constipation, is a guanylate cyclase C receptor agonist that up-regulates cyclic guanosine monophosphate (cGMP) within the cells lining the gut (enterocytes), resulting in intraluminal secretions and accelerated intestinal transit, which promotes laxation. It also has analgesic properties by reducing visceral hypersensitivity. However, there are no trials reported to date in patients with PD. Lubiprostone, 8–24 μg tablets taken twice daily, activates chloride type 2 channels (CIC-2) located on the surface of the gut epithelial cells resulting in release (efflux) of fluid into the gut lumen. Prucalopride available as 2 mg tablets is a selective serotonin (5-HT4) receptor agonist which induces a prokinetic (propulsive) effect.

There is no good evidence for the effectiveness of non-pharmacological measures for constipation in patients with PD. Nevertheless, since the reason for constipation in this population is multifactorial, a multifaceted management should be considered, ideally based on identified focal or diffuse abnormalities. Providing sufficient hydration and nutrition, ensuring privacy and comfort to allow a patient to defecate normally, the use of a commode or footstool to ensure a sitting position with knees above hips, and the availability of a caregiver to help are recommended, although, in many cases, difficult to achieve. For example, fiber-rich diet or aggressive hydration in cachectic patients who are typically bed bound is relatively contraindicated. The benefits of acupuncture and the effectiveness and safety of adjunctive traditional Chinese herbs for constipation in PD are questionable.

Physiotherapy

There is some evidence supporting abdominal massage in reducing gastrointestinal symptoms and improving bowel efficiency, particularly in those patients with concomitant neurogenic problems, such as PD. Recently, an automated intermittent colonic exo-peristalsis (ICE) treatment device significantly increased the number of complete bowel movements per week by 1.8 over baseline, reduced the symptoms of chronic constipation (Knowles Eccersley Scott Symptom Score −3.9), improved quality of life (PAC-QOL −0.7), and reduced the use of laxatives. Furthermore, treatment-related adverse events were infrequent (8.7%) and of mild to moderate intensity.

Slow transit constipation (STC) may coexist with dyssynergic defecation. In patients with gut PD with both dyssynergia and slow colonic transit successful dyssynergia treatment using biofeedback may improve transit, suggesting that outlet dysfunction may be responsible for delayed colonic transit [53].

Treatment of Constipation in PD

Constipation in PD is usually caused by a combination of poor oral intake, decreased physical activity, and the action of drugs that may slow intestinal transit time. Fecal impaction, bowel obstruction, and colonic pseudo-obstruction must be ruled out before initiating therapy for constipation. Electrolyte abnormalities and other reversible causes should be corrected first. Drugs that cause constipation should be stopped, if possible. Laxatives, with or without stool softeners, can also be used in the initial treatment of constipation. Stimulant laxatives such as bisacodyl and senna alter electrolyte transport by the intestinal mucosa and increase intestinal motor activity. If these agents are not effective, osmotic agents such as PEG can be effective at improving stool frequency and consistency [54]. Polyethylene glycol solutions (without electrolytes) are available in powder form and have been found to be effective at improving chronic constipation.

Constipation may also be associated with nausea, vomiting, anal pain, and hematochezia due to anal fissure or hemorrhoids, and it may contribute to bowel obstruction and urinary retention. Further, it causes significant psychological distress indirectly affecting patients' compliance to other PD therapies, thereby impeding treatment. Because of its chronicity and severity, it frequently requires additional therapies, thereby increasing direct and indirect healthcare costs. Contributing factors to constipation are a poor general performance status, low food and fluid intake, drugs that slow down gut motility and secretion, caregiver dependency, and limited mobility due to cancer pain and fatigue [55] (Fig. 6.7).

Despite its high prevalence in PD, constipation remains undertreated. All PD patients should be regularly assessed for constipation using patient-reported outcome measures, when feasible. Since opioids may be used for treating moderate and severe pain, they may often be a key culprit, depending on the frequency, type, and route of administration. Opioid-induced constipation (OIC) affects opioid therapy, with patients seeking a balance between pain relief and the occurrence of constipation, modifying or discontinuing their opioid use [26]. The Rome IV criteria for opioid-induced constipation (OIC) are sufficiently sensitive (85%) and specific (77%) [56, 57]. There are no significant differences between transdermal therapeutic systems and orally administered opioids [58].

Dietary and Behavioral Interventions

Antiparkinsonian and antiemetic drugs can contribute to constipation in patients with PD. The effectiveness of non-pharmacological measures for constipation in PD patients is unclear. There is potential for specific rehabilitation strategies and

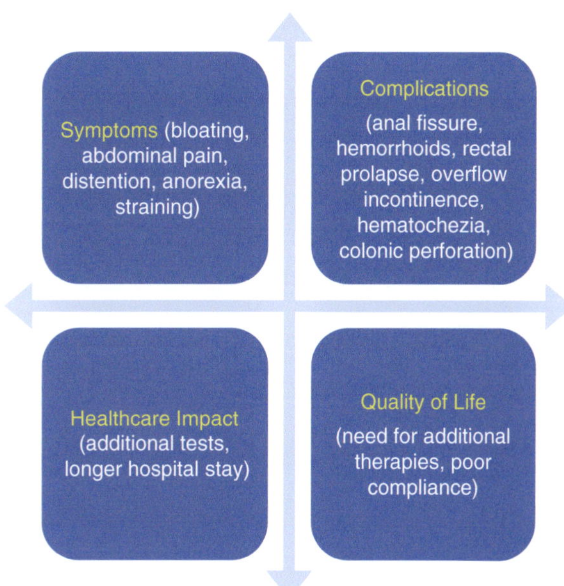

Fig. 6.7 Impact of constipation in patients with Gut PD

alternative strategies for improving constipation and minimizing the use of laxatives. Most studies show that physical exercise, abdominal massage, acupuncture, and education on the correct defecation position have a positive symptomatic impact and improvement of quality of life, but evidence for this in gut PD is indirect at best. Physiotherapy, involving massages as well as aerobic and resistance training, improves constipation particularly in women, regardless of age, sex, and frailty. The outcomes regarding TENS are inconsistent. Stimulating trigger points through acupressure and acupuncture is also increasingly used. Adequate hydration and nutrition, proper access and position for defecation using a commode or a footstool to ensure a sitting position with knees above hips, and caregiver availability help. However, fiber-rich diet or aggressive hydration is generally contraindicated [59]. In terms of non-pharmacologic interventions, there is no good evidence for acupuncture for OIC [60]. Abdominal massage may reduce symptoms and improve bowel evacuation. An automated intermittent colonic exo-peristalsis (ICE) treatment device increased the number of complete bowel movements and reduced symptoms of chronic constipation and laxative use, with infrequent mild to moderate side effects [61]. The best type, timing, and duration of treatment are unclear (Figs. 6.8, 6.9).

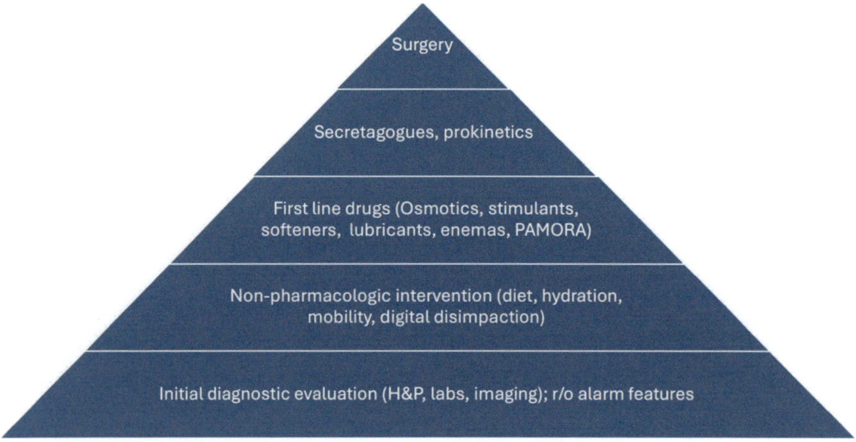

Fig. 6.8 Pyramid of management of chronic constipation in PD

Osmotic laxatives	Stimulants	Prosecretory agents	Serotonergic agents
Draw and retain water in the gut lumen	Directly increase colonic motility, reduce transit time, and increase the water content of the stool	Activate gut receptors increasing electrolyte and water secretion in the gut lumen	Activate gut receptors that increase colonic muscle contractions and peristalsis
Powder	Pills Suppositories Enemas	Pills	Pills
Polyethylene glycol	Senna Bisacodyl	Linaclotide Plecanatide Lubiprostone	Prucalopride
Cost +	Cost +	Cost ++	Cost +++

Fig. 6.9 Pharmacologic option in management of constipation in Gut PD. *Treatments may be used alone, interchangeably, or in combination

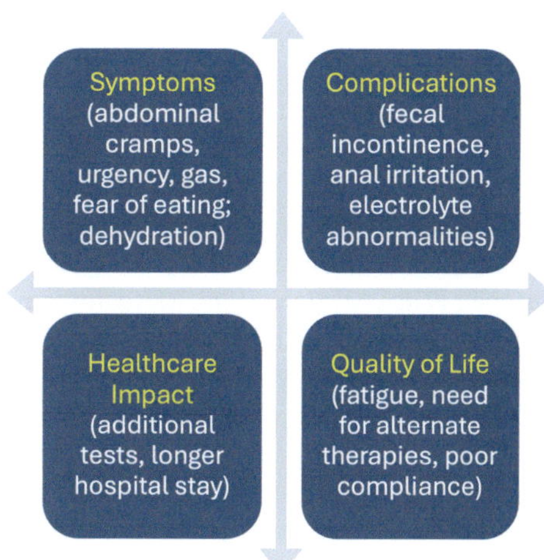

Fig. 6.10 Clinical impact of diarrhea in patients with Gut PD

Treatment of Diarrhea

Diarrhea is an uncommon event in gut PD and requires thorough investigation, particularly if it is debilitating. In severe cases, it can even be life-threatening. The onset of diarrhea can lead to treatment delays, reduced quality of life, and diminished drug compliance. The most significant cause of diarrhea is *C. difficile* infection, often a result of prior use of antibiotics, particularly in hospitalized patients (Fig. 6.10).

Drugs for Esophageal Dysmotility

There is no clear association between ineffective esophageal motility (IEM) and esophageal symptoms, and if no symptoms are present, patients could be reassured and left untreated. However, dysphagia, often to both solids and liquids, is commonly associated with IEM [62]. As there is no therapy that accelerates esophageal transit, management relies on dietary and lifestyle modification, such as cutting food into small pieces, chewing food thoroughly, eating in upright position, performing hard swallows, and consuming foods along with liquids, particularly carbonated beverages, to facilitate esophageal emptying. Medication review may help identify other drugs (i.e., anticholinergics), contributing to esophageal dysmotility.

Prokinetics, such as the serotonin-modulating agents (prucalopride or mosapride), muscarinic receptor-modulating agents (bethanechol), or motilin receptor agonists (erythromycin) have been used [63]. Mosapride may shorten esophageal transit time [64]. Metoclopramide, a dopamine-2 receptor blocker, is not effective in IEM and given its side effect profile, including risk of extrapyramidal manifestations, is not recommended in patients with PD.

While heartburn and/or regurgitation is often associated with IEM, it is unclear if they are causally related. GERD could lead to decreased peristaltic contractility, or perhaps IEM makes a patient prone to GERD because of decreased ability to clear refluxate [65]. In patients with IEM presenting with symptoms of gastroesophageal reflux and ambulatory pH testing showing pathologic esophageal acid exposure, acid suppression should be first tried [66].

In patients with persistent symptoms, prokinetics, including serotonin-modulating agents (prucalopride), muscarinic receptor-modulating agents (bethanechol), or motilin receptor agonists (erythromycin), are often considered, but their efficacy is uncertain. Metoclopramide, a dopamine-2 receptor blocker, carries the risk of extrapyramidal manifestations, which is not recommended. Neuromodulators may also play an important role to decrease symptom intensity, recognizing that these agents have no effect on peristaltic strength.

Drugs for Symptoms of Gastroparesis

Treatment of gastroparesis in PD is challenging because levodopa may further delay stomach emptying and impair the absorption of the medications used to improve the stomach symptoms. A widely used dopamine antagonist, metoclopramide, is generally contraindicated in patients with PD because it crosses into the brain and worsens the motor dysfunction. Another dopamine antagonist, domperidone, does not cross into the brain, but it carries a higher risk of causing dangerous heart irregularities. Prucalopride is a newer 5-HT4 agonist that has prokinetic effects without causing heart arrhythmias and is approved for use in chronic constipation and may benefit patients with gastroparesis but has not been well studied in PD. Azithromycin, a macrolide antibiotic, may be tried. Relamorelin, a ghrelin receptor agonist, may marginally improve gastric emptying (by a mean of 8–11 min) in idiopathic and diabetic gastroparesis, but adverse effects, mostly headaches and dizziness, are limiting its use. Endoscopically administered botulinum neurotoxin injections into the pyloric sphincter may improve symptoms of gastroparesis in patients with PD, but they need to be repeated every 6–12 months since the clinical benefit abates over time. Improvement in gastric emptying after deep brain stimulation (DBS) of the subthalamic nucleus has been reported. Although not studied in patients with PD, gastric electrical stimulation (Enterra) might be effective in refractory patients.

Prokinetics as single modality therapy has little benefit for patients with symptoms of gastroparesis. Neuromodulators as the only new therapy may decrease symptoms other than nausea and vomiting. Adding gastroparesis diets or neuromodulators to prokinetics may offer relief, suggesting that combination therapies may be more useful in managing these patients [67].

References

1. Buongiorno M, Antonelli F, Cámara A, Puente V, de Fabregues-Nebot O, Hernandez-Vara J, Calopa M, Pascual-Sedano B, Campolongo A, Valldeoriola F, Tolosa E, Kulisevsky J, Martí MJ. Long-term response to continuous duodenal infusion of levodopa/carbidopa gel in patients with advanced Parkinson disease: the Barcelona registry. Parkinsonism Relat Disord. 2015;21(8):871–6. https://doi.org/10.1016/j.parkreldis.2015.05.014. Epub 2015 May 19.
2. Almayoof M, Abbarh S, Mohammed A, Sawaf B, Alawad MJ, Alobaidy M. Esophageal ulceration following the ingestion of a single dose of doxycycline: a case report. Cureus. 2024;16(3):e57043.
3. Kim S, Suh HS. Current status of parkinsonism-related adverse events and associated drugs in Korea. J Patient Saf. 2019;15(4):e56–9.
4. Dipalma JA, Cleveland MV, McGowan J, Herrera JL. A randomized, multicenter, placebo-controlled trial of polyethylene glycol laxative for chronic treatment of chronic constipation. Am J Gastroenterol. 2007;102(7):1436–41. https://doi.org/10.1111/j.1572-0241.2007.01199.x. Epub 2007 Mar 31.
5. Corazziari E, Badiali D, Bazzocchi G, Bassotti G, Roselli P, Mastropaolo G, Lucà MG, Galeazzi R, Peruzzi E. Long term efficacy, safety, and tolerability of low daily doses of isosmotic polyethylene glycol electrolyte balanced solution (PMF-100) in the treatment of functional chronic constipation. Gut. 2000;46(4):522–6. https://doi.org/10.1136/gut.46.4.522. PMID: 10716682; PMCID: PMC1727881.
6. Lee-Robichaud H, Thomas K, Morgan J, Nelson RL. Lactulose versus polyethylene glycol for chronic constipation. Cochrane Database Syst Rev. 2010;7:CD007570. https://doi.org/10.1002/14651858.CD007570.pub2.
7. Lembo A, Camilleri M. Chronic constipation. N Engl J Med. 2003;349(14):1360–8. https://doi.org/10.1056/NEJMra020995.
8. Dupont C, Hébert G. Magnesium sulfate-rich natural mineral waters in the treatment of functional constipation—a review. Nutrients. 2020;12(7):2052. https://doi.org/10.3390/nu12072052. PMID: 32664341; PMCID: PMC7400933.
9. Dupont C, Campagne A, Constant F. Efficacy and safety of a magnesium sulfate-rich natural mineral water for patients with functional constipation. Clin Gastroenterol Hepatol. 2014;12(8):1280–7. https://doi.org/10.1016/j.cgh.2013.12.005. Epub 2013 Dec 14.
10. Johanson JF, Ueno R. Lubiprostone, a locally acting chloride channel activator, in adult patients with chronic constipation: a double-blind, placebo-controlled, dose-ranging study to evaluate efficacy and safety. Aliment Pharmacol Ther. 2007;25(11):1351–61. https://doi.org/10.1111/j.1365-2036.2007.03320.x.
11. Rao SS, Manabe N, Karasawa Y, Hasebe Y, Nozawa K, Nakajima A, Fukudo S. Comparative profiles of lubiprostone, linaclotide, and elobixibat for chronic constipation: a systematic literature review with meta-analysis and number needed to treat/harm. BMC Gastroenterol. 2024;24(1):12.

12. Layer P, Stanghellini V. Review article: Linaclotide for the management of irritable bowel syndrome with constipation. Aliment Pharmacol Ther. 2014;39(4):371–84.
13. Bassotti G, Usai-Satta P, Bellini M. Linaclotide for the treatment of chronic constipation. Expert Opin Pharmacother. 2018;19(11):1261–6. https://doi.org/10.1080/14656566.2018.1494728. Epub 2018 Jul 9.
14. Lembo AJ, Schneier HA, Shiff SJ, Kurtz CB, MacDougall JE, Jia XD, Shao JZ, Lavins BJ, Currie MG, Fitch DA, Jeglinski BI, Eng P, Fox SM, Johnston JM. Two randomized trials of linaclotide for chronic constipation. N Engl J Med. 2011;365(6):527–36. https://doi.org/10.1056/NEJMoa1010863.
15. Schoenfeld P, Lacy BE, Chey WD, Lembo AJ, Kurtz CB, Reasner DS, Bochenek W, Tripp K, Currie MG, Fox SM, Blakesley RE, O'Dea CR, Omniewski ND, Hall ML. Low-dose linaclotide (72 μg) for chronic idiopathic constipation: a 12-week, randomized, double-blind, placebo-controlled trial. Am J Gastroenterol. 2018;113(1):105–14. https://doi.org/10.1038/ajg.2017.230. Epub 2017 Aug 22. PMID: 29091082; PMCID: PMC5770595.
16. BrennerDM FR, Dorn SD, Krause R, Eng P, Kirshoff R, et al. Efficacy, safety, and tolerability of plecanatide in patients with irritable bowel syndrome with constipation: results of two phase 3 randomized clinical trials. Am J Gastroenterol. 2018;113(5):735–45.
17. Menees SB, Franklin H, Chey WD. Evaluation of Plecanatide for the treatment of chronic idiopathic constipation and irritable bowel syndrome with constipation in patients 65 years or older. Clin Ther. 2020;42(7):1406–1414.e4. https://doi.org/10.1016/j.clinthera.2020.05.012. Epub 2020 Jul 10.
18. Kamuda JA, Mazzola N. Plecanatide (trulance) for chronic idiopathic constipation and irritable bowel syndrome with constipation. P T. 2018;43(4):207–32. PMID: 29622940; PMCID: PMC5871240).
19. Bassotti G, Usai Satta P, Bellini M. Prucalopride for the treatment of constipation: a view from 2015 and beyond. Expert Rev Gastroenterol Hepatol. 2019;13(3):257–62. https://doi.org/10.1080/17474124.2019.1568238. Epub 2019 Jan 23.
20. Müller-Lissner S, Rykx A, Kerstens R, Vandeplassche L. A double-blind, placebo-controlled study of prucalopride in elderly patients with chronic constipation. Neurogastroenterol Motil. 2010;22(9):991–8. https://doi.org/10.1111/j.1365-2982.2010.01533.x. Epub 2010 Jun 7.
21. Camilleri M, Van Outryve MJ, Beyens G, Kerstens R, Robinson P, Vandeplassche L. Clinical trial: the efficacy of open-label prucalopride treatment in patients with chronic constipation—follow-up of patients from the pivotal studies. Aliment Pharmacol Ther. 2010;32(9):1113–23. https://doi.org/10.1111/j.1365-2036.2010.04455.x. Epub 2010 Sep 16.
22. Cinca R, Chera D, Gruss HJ, Halphen M. Randomised clinical trial: macrogol/PEG 3350+ electrolytes versus prucalopride in the treatment of chronic constipation—a comparison in a controlled environment. Aliment Pharmacol Ther. 2013;37(9):876–86. https://doi.org/10.1111/apt.12278. Epub 2013 Mar 11.
23. King AJ, Siegel M, He Y, Nie B, Wang J, Koo-McCoy S, et al. Inhibition of sodium/hydrogen exchanger 3 in the gastrointestinal tract by tenapanor reduces paracellular phosphate permeability. Sci Transl Med. 2018;10(456):eaam6474.
24. Soffer EE, Metcalf A, Launspach J. Misoprostol is effective treatment for patients with severe chronic constipation. Dig Dis Sci. 1994;39(5):929–33. https://doi.org/10.1007/BF02087539.
25. Taghavi SA, Shabani S, Mehramiri A, Eshraghian A, Kazemi SM, Moeini M, Hosseini-Asl SM, Saberifiroozi M, Alizade-Naeeni M, Mostaghni AA. Colchicine is effective for short-term treatment of slow transit constipation: a double-blind placebo-controlled clinical trial. Int J Color Dis. 2010;25(3):389–94. https://doi.org/10.1007/s00384-009-0794-z. Epub 2009 Aug 25.
26. Argoff CE. Opioid-induced constipation. Clin J Pain. 2020;36(9):716–22.
27. Farmer AD, Drewes AM, Chiarioni G, De Giorgio R, O'Brien T, Morlion B, Tack J. Pathophysiology and management of opioid-induced constipation: European expert consensus statement. United Eur Gastroenterol J. 2019;7(1):7–20.
28. Hanson B, Siddique SM, Scarlett Y, Sultan S. American Gastroenterological Association Institute technical review on the medical management of opioid-induced constipation. Gastroenterology. 2019;156(1):229.

29. Crockett SD, Greer KB, Heidelbaugh JJ, Falck-Ytter Y, Hanson BJ, Sultan S. American Gastroenterological Association Institute guideline on the medical management of opioid-induced constipation. Gastroenterology. 2019;156(1):218–26.
30. Bull J, Wellman CV, Israel RJ, Barrett AC, Paterson C, Forbes WP. Fixed-dose subcutaneous methylnaltrexone in patients with advanced illness and opioid-induced constipation: results of a randomized, placebo-controlled study and open-label extension. J Palliat Med. 2015;18(7):593–600.
31. Daniali M, Nikfar S, Abdollahi M. Evaluating naloxegol for the treatment of opioid-induced constipation. Expert Opin Pharmacother. 2020;21(8):883–91.
32. Esmadi M, Ahmad D, Hewlett A. Efficacy of naldemedine for the treatment of opioid-induced constipation: a meta-analysis. J Gastrointest Liver Dis. 2019;28(1):41–6.
33. Vickery R, Li Y, Schwertschlag U, Singla N, Webster L, Canafax D. With a durable increase in bowel movement frequency and return toward normal bowel function in a 5-week Ph2b opioid-induced constipation (OIC) study. J Pain. 2013;14(4):S78. https://doi.org/10.1016/j.jpain.2013.01.651.
34. Sanders M, Jones S, Löwenstein O, Jansen J-P, Miles H, Simpson K. New formulation of sustained release naloxone can reverse opioid induced constipation without compromising the desired opioid effects. Pain Med. 2015;16(8):1540–50.
35. Thor KB, Marson L, Katofiasc MA, Ricca DJ, Burgard EC. Recent developments in on-demand voiding therapies. J Pharmacol Exp Ther. 2024;390(3):302–17. https://doi.org/10.1124/jpet.123.002073. Epub ahead of print.
36. Regnard C, Twycross R, Mihalyo M, Wilcock A. Loperamide. J Pain Symptom Manag. 2011;42(2):319–23. https://doi.org/10.1016/j.jpainsymman.2011.06.001. Epub 2011 Jun 23.
37. Cann PA, Read NW, Holdsworth CD, Barends D. Role of loperamide and placebo in management of irritable bowel syndrome (IBS). Dig Dis Sci. 1984;29(3):239–47. https://doi.org/10.1007/BF01296258.
38. Ooms LA, Degryse AD, Janssen PA. Mechanisms of action of loperamide. Scand J Gastroenterol Suppl. 1984;96:145–55.
39. Efskind PS, Bernklev T, Vatn MH. A double-blind placebo-controlled trial with loperamide in irritable bowel syndrome. Scand J Gastroenterol. 1996;31(5):463–8. https://doi.org/10.3109/00365529609006766.
40. Penfold D, Volans GN. Overdose from lomotil. Br Med J. 1977;2(6099):1401–2. https://doi.org/10.1136/bmj.2.6099.1401. PMID: 589232; PMCID: PMC1632364.
41. Wedlake L, A'Hern R, Russell D, Thomas K, Walters JR, Andreyev HJ. Systematic review: the prevalence of idiopathic bile acid malabsorption as diagnosed by SeHCAT scanning in patients with diarrhoea-predominant irritable bowel syndrome. Aliment Pharmacol Ther. 2009;30(7):707–17. https://doi.org/10.1111/j.1365-2036.2009.04081.x. Epub 2009 Jun 30.
42. Lacy BE. Emerging treatments in neurogastroenterology: eluxadoline—a new therapeutic option for diarrhea-predominant IBS. Neurogastroenterol Motil. 2016;28(1):26–35.
43. Lamberts SW, van der Lely AJ, de Herder WW, Hofland LJ. Octreotide. N Engl J Med. 1996;334(4):246–54. https://doi.org/10.1056/NEJM199601253340408.
44. Dueno MI, Bai JC, Santangelo WC, Krejs GJ. Effect of somatostatin analog on water and electrolyte transport and transit time in human small bowel. Dig Dis Sci. 1987;32(10):1092–6. https://doi.org/10.1007/BF01300194.
45. Hills JM, Aaronson PI. The mechanism of action of peppermint oil on gastrointestinal smooth muscle. An analysis using patch clamp electrophysiology and isolated tissue pharmacology in rabbit and Guinea pig. Gastroenterology. 1991;101(1):55–65.
46. Pimentel M, Chang C, Chua KS, Mirocha J, DiBaise J, Rao S, Amichai M. Antibiotic treatment of constipation-predominant irritable bowel syndrome. Dig Dis Sci. 2014;59(6):1278–85. https://doi.org/10.1007/s10620-014-3157-8. Epub 2014 May 1.
47. Hanna-Jairala I, Drossman DA. Central neuromodulators in irritable bowel syndrome: why, how, and when. Am J Gastroenterol. 2024;119(7):1272–84.
48. Müller-Lissner SA, Kamm MA, Scarpignato C, Wald A. Myths and misconceptions about chronic constipation. Am J Gastroenterol. 2005;100:232–42.

49. Bassotti G, Gaburri M. Manometric investigation of high-amplitude propagated contractile activity of the human colon. Am J Phys. 1988;255(5 Pt 1):G660–4.
50. Stephen AM, Cummings JH. Mechanism of action of dietary fibre in the human colon. Nature. 1980;284(5753):283–4. https://doi.org/10.1038/284283a0.
51. Rao SS, Yu S, Fedewa A. Systematic review: dietary fibre and FODMAP-restricted diet in the management of constipation and irritable bowel syndrome [Erratum in: Aliment Pharmacol Ther. 2015 Aug;42(4):490.]. Aliment Pharmacol Ther. 2015;41(12):1256–70. https://doi.org/10.1111/apt.13167. Epub 2015 Apr 22.
52. Bharucha AE, Pemberton JH, Locke GR 3rd. American Gastroenterological Association technical review on constipation. Gastroenterology. 2013;144(1):218–38. https://doi.org/10.1053/j.gastro.2012.10.028. PMID: 23261065; PMCID: PMC3531555.
53. Rao SS, Mudipalli RS, Stessman M, Zimmerman B. Investigation of the utility of colorectal function tests and Rome II criteria in dyssynergic defecation (Anismus). Neurogastroenterol Motil. 2004;16(5):589–96. https://doi.org/10.1111/j.1365-2982.2004.00526.x.
54. Brandt LJ, Prather CM, Quigley EM, Schiller LR, Schoenfeld P, Talley NJ. Systematic review on the management of chronic constipation in North America. Am J Gastroenterol. 2005;100(Suppl 1):S5–S21.
55. Hess B, Bernardi M, Klotz HP. Attitude of Swiss physicians towards opioid-induced constipation: a national survey. Eur J Intern Med. 2011;22(5):527–31. https://doi.org/10.1016/j.ejim.2011.02.014. Epub 2011 Mar 21.
56. Lacy BE, Mearin F, Chang L, Chey WD, Lembo AJ, Simren M, Spiller R. Bowel disorders. Gastroenterology. 2016;150(6):1393–140.
57. Davies A, Leach C, Butler C, Gregory A, Henshaw S, Minton O, Shorthose K, Batsari KM. Opioid-induced constipation in patients with cancer: a "real-world," multi-centre, observational study of diagnostic criteria and clinical features. Pain. 2021;162(1):309–18.
58. Ueberall MA, Mueller-Schwefe G. Opioid-induced constipation—a frequent and distressing side effect in daily practice affecting oral and transdermal opioid applications. Eur J Pain. 2006;10(S1):S172.
59. Crawford GB, Dzierżanowski T, Hauser K, Larkin P, Luque-Blanco AI, Murphy I, Puchalski CM, Ripamonti CI. Care of the adult cancer patient at the end of life: ESMO clinical practice guidelines. ESMO Open. 2021;6(4):100225. https://doi.org/10.1016/j.esmoop.2021.100225.
60. Yang P, Wang Y, Xiao Y, Ma Q, Ma R, Mi J, Hui J. Acupuncture for opioid-induced constipation. Medicine (Baltimore). 2020;99(49):e23352. https://doi.org/10.1097/MD.0000000000023352.
61. McClurg D, Booth L, Herrero-Fresneda I. Safety and efficacy of intermittent colonic exoperistalsis device to treat chronic constipation: a prospective multicentric clinical trial. Clin Transl Gastroenterol. 2020;11(12):e00267. https://doi.org/10.14309/ctg.0000000000000267.
62. Kamboj AK, Katzka DA, Vela MF, Yadlapati R, Ravi K. A practical approach to ineffective esophageal motility. Neurogastroenterol Motil. 2024;4:e14839. https://doi.org/10.1111/nmo.14839. Epub ahead of print.
63. Jandee S, Geeraerts A, Geysen H, Rommel N, Tack J, Vanuytsel T. Management of ineffective esophageal hypomotility. Front Pharmacol. 2021;12:638915.
64. Cho YK, Choi MG, Han HW, et al. The effect of mosapride on esophageal motility and bolus transit in asymptomatic volunteers. J Clin Gastroenterol. 2006;40:286–92.
65. Dao HV, Hoang LB, Luu MHT, et al. High prevalence of esophagitis in patients with severe ineffective esophageal motility: need for a new diagnostic cutoff. Ann Gastroenterol. 2022;35:483–8.
66. Kamal AN, Kathpalia P, Leiman DA, et al. Quality indicator development for the approach to ineffective esophageal motility. J Clin Gastroenterol. 2024;58:975. https://doi.org/10.1097/MCG.0000000000001963.
67. Hasler WL, Lee AA, Moshiree B, Surjanhata BC, Rao S, Parkman HP, Nguyen LA, Sarosiek I, Wo JM, Schulman MI, McCallum RW, Kuo B. Benefits of prokinetics, gastroparesis diet, or neuromodulators alone or in combination for symptoms of gastroparesis. Clin Gastroenterol Hepatol. 2024;22(4):867–877.e12. https://doi.org/10.1016/j.cgh.2023.10.014. Epub 2023 Oct 30.

7 Non-pharmacologic Interventions in Gut PD

Non-pharmacologic interventions in gut PD are commonly used either alone or in conjunction with various other therapies in PD. Multidisciplinary discussion and consideration of all available options and their advantages or disadvantages as well as their potential to interfere with ongoing therapies implemented by the movement disorder specialist is essential as part of the overall strategic management of the disease. After such assessment, these non-pharmacologic interventions may be initiated and implemented by the patients themselves (i.e., diet and exercise), by specialists (i.e., swallowing rehabilitation or gut-directed pelvic floor behavioral therapy), or by gastroenterologists (i.e., Botox injection therapies or balloon dilation). It is also important to keep in mind that there is inadequate evidence that these interventions are widely effective, since experience is limited in small studies without adequate controls or well-accepted criteria set for defining success. Yet, since such trials are difficult to implement, patients are advised to explore these options in consultation with their clinical team, depending on the availability of the local expertise. The impact of some of these interventions may vary, depending on disease severity and evolution, patient adherence, and operator's expertise. In some cases, the clinical impact may be transient, and a repeat effort is needed. Therefore, repeat considerations or reimplementation of therapy may need to be explored at least yearly (Fig. 7.1).

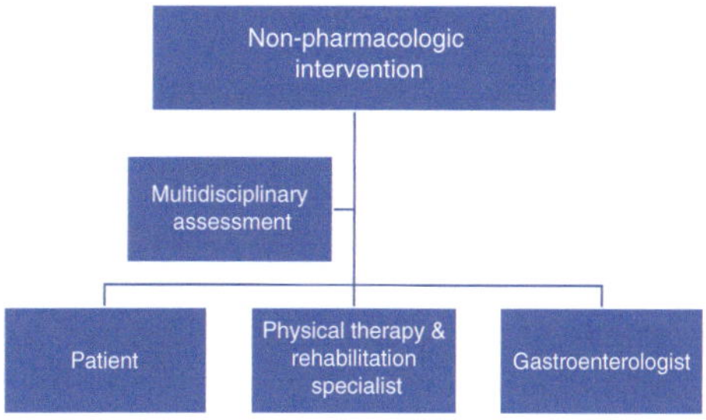

Fig. 7.1 Non-pharmacologic interventions in Gut PD require coordination among patients and various providers

G. Triadafilopoulos, *The Gut in Parkinson's Disease*,
https://doi.org/10.1007/978-3-031-77171-2_7

Diet and Lifestyle in Gut PD

Regular exercise, a healthy diet, high-quality sleep, and avoidance of adverse exposures have been associated with reduced mortality and provide a foundation for advising persons with Parkinson's disease at any stage [1]. The Mediterranean diet is the most advisable diet for gut PD. It consists mostly of plant foods (fruits, vegetables, cereals, nuts, and seeds), with moderate amounts of dairy products, mostly fermented (e.g., cheese and yogurt); low-to-moderate amounts of fish and poultry; low amounts of red meat; and usually wine along with the meal [2]. There is robust evidence for a protective association between adherence to the Mediterranean diet and neurodegenerative diseases [3]. The beneficial effects of this diet may be mediated through several mechanisms, such as reductions of blood lipids and inflammatory and oxidative stress markers, enhancement of endothelial and antithrombotic function, and a reduction in cerebral neurodegenerative changes [4].

The nutrient capacity of a low-fat diet where lipids contribute less than 30% of total energy intake depends on individual food choices, but it may be better tolerated by patients with gastroparesis [5]. The more restrictive the diet, the higher the risk of various nutrient deficiencies [6]. The MIND (Mediterranean–DASH Intervention for Neurodegenerative Delay) is a dietary approach to improve cognitive function, and it emphasizes consumption of whole grains, vegetables, beans, and nuts, with a unique focus on berries and green, leafy vegetables [7], while it restricts consumption of red meat, as well as other foods with high total and saturated lipid content (fast and fried foods, cheese, butter and margarine, and pastries and sweets), while it maintains olive oil as the primary oil. Consumption of fish more than once a week and poultry more than twice a week is recommended [8].

Adherence to a Mediterranean diet was associated with a reduced risk of PD in large US and Swedish cohort studies [9, 10] and a reduced risk of prodromal PD among older adults in Greece [11]. The combination of the Mediterranean and the DASH (Dietary Approaches to Stop Hypertension) diet was associated with later age at PD onset in women but not in men [12].

There is no association between dietary intake of macronutrients (carbohydrate, protein, fat) or total calories and Parkinson's disease (PD) risk [13]. However, there may be beneficial effects of polyunsaturated fatty acid intake that plays a key role in inflammatory processes [14].

Investigations into micronutrients have focused on dietary antioxidants (vitamins C and E and carotenoids). A meta-analysis concluded that high vitamin E intake was associated with reduced PD risk [15]. Dietary vitamin C and carotenoids are not helpful. A reduced PD risk has been linked to vitamin B6 intake [16] but not that of vitamin B12 or folate [17]. In longitudinal studies of people with PD, higher homocysteine levels predicted worse motor and cognitive decline [18]. Limited studies on the relationship between dietary minerals and PD risk have found no association between iron, zinc, or copper intake and PD risk [19].

There is a dose-dependent inverse relationship between caffeine consumption and PD risk [20, 21]. Coffee and its compounds (including polyphenols) also modulate the gut microbiome and promote gut motility [22]. Evidence of disease-modifying effects of caffeine in people diagnosed with PD is mixed. Consumption of tea, which is rich in flavonoids, has been linked to reduced PD risk, although the effects could be partly attributable to caffeine. Dietary flavonoids in fruits, fruit juices, and red wine are associated with reduced PD risk [23].

Higher plasma urate levels have been linked to reduced risk and slower progression of PD in several prospective studies [24], and one study found that dietary factors predictive of higher plasma urate levels were associated with reduced PD risk [25]. There is no association between total dairy product consumption and dietary calcium and PD risk [26]. Among specific dairy products, higher milk consumption (in particular, low- fat milk) seemed to be associated with a higher risk of PD [27]. No consistent association was found with other dairy products, such as yogurt, cheese, ice cream, or butter [26]. The link between alcohol consumption and PD risk is uncertain [28]. Possible biological mechanisms linking alcohol and PD include antioxidant effects through urate-elevating effects of beer or polyphenols in wine [29].

Nonprescription Therapeutics

Peppermint oil is well known to inhibit colonic peristalsis. Its primary active component is L-menthol, which has anti-inflammatory, antispasmodic, and analgesic properties [30]. The antispasmodic effect appears to be dependent on the direct uptake by the gastrointestinal mucosa [31]. Orally administered peppermint oil has been shown to be safe with no serious adverse effects. A novel triple-coated microsphere formulation of purified L-menthol designed for sustained release in the small intestine marketed as OTC IBGard™ seems effective at reducing symptoms with minimal side effects [32]. IBGard™ can be taken daily and is clinically shown to help manage a combination of occasional symptoms, such as cramping, bowel urgency, and diarrhea within 24 h.

Swallowing Rehabilitation

Training in expiratory muscle strength, various postural techniques, oral motor exercises, video-assisted swallowing therapy, surface electrical stimulation, thermal stimulation, touch, compensatory interventions, training regime for swallowing, neuromuscular electrical stimulation, Lee Silverman voice treatment, swallow maneuver, airway protection, and postural compensation maneuvers have been used, based on the results of formal assessment by speech therapy. Several rehabilitative therapies have been successful, with significant improvement seen in the degenerative function (coordination, speed, and volume), quality of life, and social relationships of people with PD [33].

Gut-Directed Pelvic Floor Behavioral Therapy (PFBT) for Constipation and Fecal Incontinence

Evacuation disorder (ED) implies inability to evacuate stool due to impaired coordination of the abdominal, recto-anal, and puborectalis muscles. ED is common in patients with gut PD and negatively impacts quality of life and may cause anxiety and/or depression [34]. The diagnosis of ED requires abnormal results of at least two of three anorectal physiologic tests (HR-ARM, RST, BET) or dynamic imaging [35]. Treatment of ED is best achieved by pelvic floor retraining and biofeedback therapy [36]. Although biofeedback demonstrates up to 80% efficacy in clinical trials, its labor-intensive nature, need for repetitive visits, and limited availability present challenges to accessibility, particularly in patients with PD [37]. Therefore, it is crucial to identify options that are both effective, practical for the patient, and highly accessible. The primary aim of pelvic floor behavioral therapy is to normalize bowel function by establishing regular and effective evacuation and control involuntary passage of gas and/or stool.

Fecal incontinence is the involuntary loss of stool or, rarely, gas. Many patients continue to have increased bowel frequency, urgency, fecal incontinence, constipation (low bowel frequency or difficult rectal emptying), abdominal pain, bloating, straining, and anorectal pain despite drug- or deep brain stimulation (DBS)-induced PD remission. There are many reasons for such persistent bowel symptoms, involving both physiological and psychological factors, such as alterations in rectal compliance (stool accommodation), sensitivity (ability to feel the presence of stool in the rectum), and contractility (strength of propulsion), together with anal sphincter and pelvic floor muscle dysfunction. Fear of incontinence or frequent strong urges to defecate can lead to overactive or hypervigilant pelvic floor muscle activity, which over time becomes behaviorally ingrained with patients who lose their ability to relax appropriately. Psychological factors further aggravate such dysfunction, through direct motor and sensory effects, and patients' perception of symptoms.

Persistent fecal incontinence or constipation can be treated with specific drug therapies, such as laxatives and antidiarrheal drugs, although these are often insufficient to relieve symptoms. Other therapeutic physical, psychological, or practical management strategies, such as behavioral treatment, may therefore be beneficial. Behavioral treatment comprises a combination of elements, including pelvic floor training, which is not restricted to "biofeedback" and is tailored to the individual symptoms, but it requires a mentally and physically capable and cooperative patient [38].

Pelvic floor muscle training, with or without biofeedback, is effective in managing fecal urgency, leakage, straining, incomplete evacuation, abdominal pain, and bloating. Patients should be referred by gastroenterologists to a specialist pelvic floor physiotherapist after having been studied thoroughly. All patients should have no active endoscopic disease, such as cancer, colitis, or diverticulosis. The pelvic floor therapist assesses anorectal function by digital palpation and with a rectal balloon. Assessment includes pelvic floor and anal sphincter muscle resting tone, pelvic floor and anal sphincter muscle contraction strength and endurance, ability to relax the pelvic floor and anal sphincter muscles, rectal sensation, and the ability to expel a rectal balloon. During simulated defecation (testing balloon expulsion), the therapist assesses the ability to relax pelvic floor voluntary muscles (puborectalis and external anal sphincter), maintain relaxation, and generate an effective expulsive effort [39].

Patients are deemed to have pelvic floor and anal sphincter muscle dysfunction by digital assessment using the following criteria: (1) absent or weak voluntary muscle contraction, (2) absent or partial voluntary relaxation, and (3) absent or partial involuntary relaxation during simulated defecation. Assessment of pelvic floor relaxation is subjective and variable; however, it remains a useful focus for therapy.

Patients are educated about the anatomy and function of the gut and muscles contributing to its function: the diaphragm, abdominal wall, and pelvic floor. The interaction between psychological and cognitive function and the gut and pelvic floor is also explained and referred to throughout the treatment to assist patients in understanding the potential impact of past and existing physical, psychological, and social factors on gut function. Lifestyle modifications include regularized toileting and eating habits, food and fluid intake, stress management, and general exercise.

Bowel habit training encourages defecation at a predictable time, including routine toilet sits after meals. Frequent or prolonged defecatory attempts and straining are discouraged. Muscle dysfunction is addressed through exercises designed to improve the strength, endurance, relaxation, and coordination of the pelvic floor, anal sphincter, abdominal muscles, and diaphragm. Patients learn to coordinate effective evacuation and defer defecation, if indicated.

Training is assisted by various forms of "feedback," including verbal instruction, the therapist monitoring pelvic floor contraction and relaxation using digital palpation, and the use of a rectal balloon catheter to provide a rectal sensory input. An inflated rectal balloon is used to practice simulated defecation, using therapist assistance initially until the patient can expel the balloon unassisted. Progressive balloon inflations are used to practice urge resistance at increasing volumes. Individualized home exercise programs are then prescribed to reinforce and consolidate relearned bowel behavior. Visits occur at intervals of 4 weeks. Treatment is concluded when patients are satisfied with the progress they have made, or up to five sessions.

Vibrating Capsule for Chronic Constipation

The vibrating capsule is a non-pharmacologic, orally ingested, programmable capsule that vibrates intraluminally to induce bowel movements. In a double-blind, placebo-controlled trial of 312 patients with chronic constipation, who were randomized to receive either a vibrating or placebo capsule, once daily, 5 days a week for 8 weeks, a greater percentage of patients receiving the vibrating capsule achieved both primary efficacy end points compared with placebo (39.3% vs. 22.1%, $P = 0.001$ for $CSBM_1$; 22.7% vs. 11.4% $P = 0.008$ for $CSBM_2$). Significantly greater improvements were also seen with the vibrating capsule for stool consistency and quality-of-life measures compared with placebo. A mild vibrating sensation was reported by 11% of patients in the vibrating capsule group [40]. There are no studies on the efficacy of the vibrating capsule in PD patients.

Footstool for Constipation

Body position during defecation influences stool evacuation, and squatting, as compared to the seated position, is the most common adopted position for defecation in many cultures globally. It is often recommended to improve the mechanics of defecation by increasing hip flexion, thereby relaxing the puborectalis muscle, and straightening the anorectal angle to decrease resistance during evacuation. In recent times, there has been a rise in popularity of footstools that aim to replicate the biomechanics of squatting while seated at the toilet. While a few case series suggested that using a footstool in those with abnormal BET would facilitate balloon expulsion, the best evidence comes from a recent study in which a footstool led to successful balloon evacuation in 53 out of 325 patients (16.3%) who had initially failed to evacuate the balloon within 2 min [41]. In another study of the 667 patients with DDs, a total of 251 (38%) had failed BET. A footstool corrected BET in 41 (16%) of those with failed BET. Gender-specific differences were noted in anorectal pressures, among patients with and without normal BET, revealing gender-based nuances in pathophysiology. Comparing patients who passed BET with footstool with those who did not, the presence of optimal stool consistency, with reduced instances of loose stools and decreased reliance on laxatives, was significant. Additionally, in women who benefited from a footstool, lower anal pressures at rest and simulated defecation were observed. Independent factors associated with a successful BET with a footstool in women included age <50, Bristol 3 or 4 stool consistency, lower anal resting pressure, and higher recto-anal pressure gradient [42].

Programmed Defecation

There are two types of programmed defecation that apply to patients with gut PD who have failed managing their constipation and suffer from incomplete evacuation, fecal impaction, and possibly overflow incontinence. The two types are either (1) patient-implemented or (2) assistant-implemented. The first is typically applicable to patients who do not use assistance for bowel evacuation. In such cases, the patient decides to sit at the toilet at a predicted time of the day, preferably 30 min after a meal that stimulates the gastrocolic reflex and enhances the probability for a successful stool expulsion through an enhanced colorectal contractility. Alternatively, the patient may choose to use a suppository, an enema, a footstool, or even digital maneuvers (rectal or transvaginal) to evacuate. The advantages of this approach are that a consistent effort is made by the patient to evacuate and, in this process, decide about the overall efficacy of their bowel program and potentially take alternative action. Further, the approach serves as an early notice to the development of fecal impaction for which more immediate action will be needed.

The second type of programmed defecation, the assistant-implemented, is reserved for those patients with gut PD who are much more caretaker-dependent and unable to use the bathroom on their own, safely and effectively, mostly because of frailty and poor motor coordination. Under the circumstances, the safest and most effective bowel evacuation strategy is to use digital stimulation with or without laxative suppository use, followed by digital rectal evacuation that completely empties the rectal vault of fecal contents. This is best accomplished at the bedside, with the patient turned in the left lateral position, right knee bent, adequate bedding covers, multiple glove exchanges, and proper lubrication of the anal canal and with the cooperation of the patient who will guide the assistant in their efforts aiming to avoid anal and rectal discomfort during this process that may require 30 min or longer. If this effort is made regularly—preferably daily—it mitigates the chances of fecal impaction and fecal incontinence for at least 24 h. However, the process is unpleasant, time-consuming, and messy. More importantly, it requires a trained and careful operator to minimize any local injury. The process has the added advantage of allowing more careful inspection of the perianal area (i.e., for fissures or hemorrhoids) and the posterior pelvis (i.e., for decubitus ulcers) that can be treated early and more effectively.

Endoscopic Therapies

Bleeding

Endoscopic therapy has replaced surgery as the primary intervention for gastrointestinal bleeding using a variety of effective endoscopic approaches: thermal, mechanical, and injection of pharmacologic solutions that have either biologic or mechanical properties. Of these, injection therapy is often used because of its simplicity, portability, and low cost. The mechanisms of action vary, from vascular tamponade, constriction of blood vessels, platelet aggregation, thrombosis, and vascular sclerosis [43].

Botox Injection

Normal functioning of the smooth muscles of the gut depends on a complex interaction among the parasympathetic, sympathetic, and enteric nervous systems. Often, acetylcholine acts as the presynaptic neurotransmitter to regulate smooth muscle tone. Many diseases, such as cricopharyngeal dysphagia, distal esophageal spasm, esophagogastric junction outflow obstruction (EGJOO), delayed gastric emptying, sphincter of Oddi dysfunction (SOD), chronic anal fissure, and anismus, occur because of poorly regulated tone, relaxation, and coordination. Botulinum toxin type A (Botox; Allergan, Irvine, Calif) has been used in different spastic disorders of the striated muscles when injected locally, but it can also be effective in treating smooth muscle gut disorders in PD [44].

Botulinum toxin is produced by the bacterium *Clostridium botulinum.* It selectively interacts with cholinergic nerve endings along the gut and inhibits the release of acetylcholine paralyzing the affected muscle. Although botulinum toxin irreversibly binds with such nerve terminals, sprouting terminal axons form new synapses and render the neuromuscular blockade temporary (6–12 months) [45].

Cricopharyngeal Dysphagia

The cricopharyngeal muscle separates the hypopharynx from the esophagus. Dysfunction of the cricopharyngeal muscle leads to dysphagia, regurgitation, and pulmonary aspiration, and it is common in PD. Surgical myotomy, mechanical (balloon) dilation, and neurectomy of the oropharyngeal plexus have been used, but they are invasive [46]. The safety and efficacy of botulinum toxin injection of the cricopharyngeal muscle may vary based on patient characteristics, injection technique, and the type and amount of botulinum toxin used. Most patients have been treated, under general anesthesia, by otolaryngologists with the use of a rigid laryngoscope or have been treated with transcervical injections guided by electromyography (EMG) monitoring or CT visualization. The success rate of botulinum toxin injection into the upper esophageal sphincter (UES) is 70–100%, with each injection lasting up to 14 months. For injection into the UES using a flexible endoscope, 25–100 U botulinum toxin is injected in equally divided doses into three or four quadrants of the UES [47, 48].

Esophagogastric Junction Outflow Obstruction (EGJOO)

Patients with EGJOO in the context of PD may complain of esophageal dysphagia, regurgitation, aspiration, weight loss, and chest pain. Diagnosis is made when there is incomplete relaxation of the LES on HRM. The goal of therapy is to reduce LES tone, thereby palliating patient symptoms [49]. Once the diagnosis of EGJOO is established, secondary causes need to be ruled out, especially the use of opioids. Since many patients improve without intervention, expectant management is recommended for patients with mild symptoms. Medical treatment has limited role. Botox injection into the lower esophageal sphincter is often used to see if the patient improves before committing to more definitive treatments, such as pneumatic dilation, peroral endoscopic myotomy, or Heller myotomy [50]. Pneumatic dilation has initial response rates of 60–90%, with many maintaining symptomatic relief at 5 years. Since pneumatic balloon dilation carries a significant risk of perforation, it may not be suitable for elderly patients.

For the flexible endoscopic Botox injection, 100 U botulinum toxin is injected into each of the four quadrants around the circumference of the LES, by using a 23-gauge sclerotherapy needle. With symptomatic response, there is a corresponding decrease in the sphincter pressure by up to 60%. After treatment, patients can be discharged, and they can eat, as tolerated. The main drawback of botulinum toxin therapy is the need for repeat treatments, because only one third of patients will be free of recurrent symptoms after 1 year from the first injection. The mean time until the need for repeat treatment ranges from 4 to 15 months. In responsive patients, repeat injection of 100 U may be done, as needed, to maintain prolonged clinical remission. The interval between injections appears to decrease with each additional injection. This phenomenon appears to be related to antibody formation in patients with PD who also receive Botox for skeletal muscle spasticity and rigidity.

Pyloric Botox Injection

As a potent inhibitor of neuromuscular transmission, botulinum toxin may be used to treat patients with symptoms of PD-associated gastroparesis yielding mild improvement in gastric emptying time and a modest improvement in symptoms [51]. Better clinical response is usually seen with larger (200 U) doses, particularly in PD patients, temporizing their symptoms and improving levodopa pharmacodynamics for several months [52]. It may also be used to assess response prior to pyloromyotomy or G-POEM: a positive clinical response to pyloric Botox injection may suggest better response [53].

Pyloric Balloon Dilation

Pyloric dilation may be performed during an upper gastrointestinal endoscopy using a hydraulic through-the-scope balloon placed through the pylorus, preferably over a guidewire, under endoscopic guidance. The balloon is then inflated with water for 1 min at a diameter of 15 mm, for 1 min at a diameter of 16.5 mm, and then for 3 min at a diameter of 18 mm. The balloon is then removed. Immediate complications such as perforation or bleeding are rare. Data on the efficacy of pyloric dilation in refractory gastroparesis from gut PD are scarce. Over half of patients with gastroparesis respond to pyloric balloon dilation at 2 months, and up to 30–40% are symptom-free at 6–24 months [54]. For refractory patients with gut PD, we have also used a combination approach of pyloric Botox injection and dilation, at the same endoscopy setting.

The severity of gastroparesis and its response to various interventions is typically assessed by the GCSI score [55], or, practically, using a Likert scale. The GCSI score assesses nine items: nausea, regurgitation, vomiting, gastric fullness, early satiety, postprandial fullness, loss of appetite, bloating, and abdominal distension, each rated on a scale from 0 to 5, with higher scores revealing a more symptomatic disease. The final score is also between 0 and 5. In the GCSI, vomiting is assessed by the number of episodes per week. Epigastric pain is assessed in a binary way, by the presence of intense pain at least once a day.

Anal Botox Injection

Dyssynergic defecation is characterized by inappropriate coordination of the pelvic floor muscles during defecation, resulting in impaired stool expulsion. The mainstay of treatment is biofeedback, and alternative therapies are limited in those who do not respond. Botulinum toxin type A injection as a treatment option for dyssynergia in gut PD has been explored in small studies but with good results [52]. The procedure is safe and well tolerated and leads to symptomatic improvement that lasts up to several months, and it can be repeated as needed and combined with other therapies [56].

Manual or Endoscopic Stool Disimpaction

Despite the prevalence of constipation and fecal impaction in gut PD, it is shocking that a proper rectal examination is not performed more frequently. This may be due to patients' underreporting of constipation, the misconception that a bowel evacuation was indeed complete, or other factors, such as patients' discomfort or inconvenience. A rectal examination can be easily performed during an office visit or by the patient's bedside and should be considered as an important tool in the periodic assessment of patients with gut PD. If fecal impaction is seen, proper disimpaction, manually or assisted by colonoscopy, should be considered, and it has the potential of improving symptoms and quality of life as well as reducing emergency department visits and hospitalizations with their associated morbidity and mortality [57].

Bypassing the Gastrointestinal Tract

Gastrointestinal dysfunction plays a pivotal role in the development of motor fluctuations, tremor, and dyskinesia in PD. Hence, treatments that partly or completely bypass the gastrointestinal tract have been developed. Because absorption of levodopa in tablet form is affected by unpredictable gastric emptying time, many liquid levodopa formulations have become commercially available options. An alternative is orally dissolving levodopa, which is swallowed with saliva. Yet, levodopa absorption still depends on the function of the proximal small intestine and thus requires an intact and well-functioning gut lining. The stomach can be bypassed using a percutaneous endoscopic gastrostomy (PEG) with jejunal tube extension for levodopa infusion deep into the small intestine (Duopa therapy). Particularly for patients with severe swallowing problems, the use of rotigotine transdermal patch and apomorphine subcutaneous infusion or injection are options to be considered. Unfortunately, monotherapy with dopamine agonists often does not suffice as the disease and its motor symptoms progress. Thus, novel levodopa formulations are already under development: microspheres for drug delivery and esters or alkyl esters for transdermal, intranasal, subcutaneous, or intramuscular administration and micro-enemas for rectal administration.

Fecal Microbiota Transplantation (FMT)

Clinically, FMT relieved intractable constipation in a PD patient who presented with 7 years of resting tremor and bradykinesia. Multiple clinical trials are now in progress to study the potential of FMT in restoring gut homeostasis and reducing constipation and motor symptoms. Donor fecal microbiota transplantation is typically administered at colonoscopy without major adverse events. Available data from human studies are scarce, and long-term, double-blind, randomized, and controlled clinical trials are required to find the clinical efficacy and safety of these therapeutic modalities in PD patients [58]. As new evidence becomes available, FMT clinical outcomes will be refined and better assessed, both in the short and long term. Since the small intestinal microbiome is volatile and subject to fluctuations, an inadequate clinical response to FMT or symptom recurrence after FMT should not necessarily be seen as a failure. A repeat FMT alone, or in conjunction with a course of antibiotics or prebiotics, may be all that is necessary to reestablish the patient's clinical stability or improvement (Fig. 7.2).

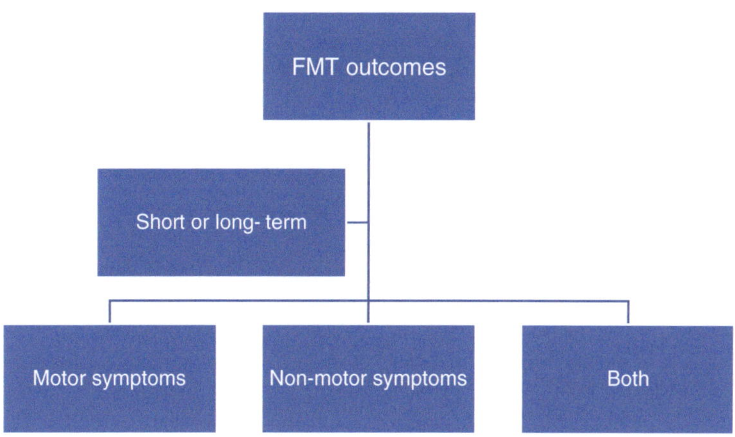

Fig. 7.2 Possible clinical outcomes after Fecal Microbiota Transfer (FMT) in patients with Gut PD

References

1. Zhang X, Molsberry SA, Schwarzschild MA, Ascherio A, Gao X. Association of diet and physical activity with all-cause mortality among adults with Parkinson disease. JAMA Netw Open. 2022;5(8):e2227738.
2. Yannakoulia M, Kontogianni M, Scarmeas N. Cognitive health and Mediterranean diet: just diet or lifestyle pattern? Ageing Res Rev. 2015;20:74–8.
3. Dinu M, Pagliai G, Casini A, Sofi F. Mediterranean diet and multiple health outcomes: an umbrella review of meta-analyses of observational studies and randomised trials. Eur J Clin Nutr. 2018;72:30–43.
4. Rainey-Smith SR, Gu Y, Gardener SL, et al. Mediterranean diet adherence and rate of cerebral Aβ-amyloid accumulation: data from the Australian imaging, biomarkers and lifestyle study of ageing. Transl Psychiatry. 2018;8:238.
5. Dunn-Emke SR, Weidner G, Pettengill EB, Marlin RO, Chi C, Ornish DM. Nutrient adequacy of a very low-fat vegan diet. J Am Diet Assoc. 2005;105:1442–6.
6. Herpich C, Müller-Werdan U, Norman K. Role of plant-based diets in promoting health and longevity. Maturitas. 2022;165:47–51.
7. Morris MC, Tangney CC, Wang Y, Sacks FM, Bennett DA, Aggarwal NT. MIND diet associated with reduced incidence of Alzheimer's disease. Alzheimers Dement. 2015;11:1007–14.
8. Scarmeas N, Anastasiou CA, Yannakoulia M. Nutrition and prevention of cognitive impairment. Lancet Neurol. 2018;17:1006–15.
9. Gao X, et al. Prospective study of dietary pattern and risk of Parkinson disease. Am J Clin Nutr. 2007;86:1486–94.
10. Yin W, Lof M, Pedersen NL, Sandin S, Fang F. Mediterranean dietary pattern at middle age and risk of Parkinson's disease: a Swedish cohort study. Mov Disord. 2021;36:255–60.
11. Maraki MI, et al. Mediterranean diet adherence is related to reduced probability of prodromal Parkinson's disease. Mov Disord. 2019;34:48–57.
12. Metcalfe- Roach A, et al. MIND and Mediterranean diets associated with later onset of Parkinson's disease. Mov Disord. 2021;36:977–84.
13. Wang A, Lin Y, Wu Y, Zhang D. Macronutrients intake and risk of Parkinson's disease: a meta-analysis. Geriatr Gerontol Int. 2015;15:606–16.
14. Marion-Letellier R, Savoye G, Ghosh S. Polyunsaturated fatty acids and inflammation. IUBMB Life. 2015;67:659–67.
15. Chang MC, Kwak SG, Kwak S. Effect of dietary vitamins C and E on the risk of Parkinson's disease: a meta-analysis. Clin Nutr. 2021;40:3922–30.
16. de Lau LML, Koudstaal PJ, Witteman JC, Hofman A, Breteler MM. Dietary folate, vitamin B12, and vitamin B6 and the risk of Parkinson disease. Neurology. 2006;67:315–8.
17. Chen H, et al. Folate intake and risk of Parkinson's disease. Am J Epidemiol. 2004;160:368–75.
18. Christine CW, et al. Vitamin B12 and homocysteine levels predict different outcomes in early Parkinson's disease. Mov Disord. 2018;33:762–70.
19. Cheng P, et al. Dietary intake of iron, zinc, copper, and risk of Parkinson's disease: a meta-analysis. Neurol Sci. 2015;36:2269–75.
20. Palacios N, et al. Caffeine and risk of Parkinson's disease in a large cohort of men and women. Mov Disord. 2012;27:1276–82.
21. Qi H, Li S. Dose-response meta-analysis on coffee, tea and caffeine consumption with risk of Parkinson's disease. Geriatr Gerontol Int. 2014;14:430–9.
22. Iriondo-DeHond A, Uranga JA, Del Castillo MD, Abalo R. Effects of coffee and its components on the gastrointestinal tract and the brain–gut axis. Nutrients. 2020;13:88.
23. Gao X, Cassidy A, Schwarzschild MA, Rimm EB, Ascherio A. Habitual intake of dietary flavonoids and risk of Parkinson disease. Neurology. 2012;78:1138–45.
24. Gao X, O'Reilly ÉJ, Schwarzschild MA, Ascherio A. Prospective study of plasma urate and risk of Parkinson disease in men and women. Neurology. 2016;86:520–6.
25. Gao X, et al. Diet, urate, and Parkinson's disease risk in men. Am J Epidemiol. 2008;167:831–8.

26. Hughes KC, et al. Intake of dairy foods and risk of Parkinson disease. Neurology. 2017;89:46–52.
27. Jiang W, Ju C, Jiang H, Zhang D. Dairy foods intake and risk of Parkinson's disease: a dose response meta-analysis of prospective cohort studies. Eur J Epidemiol. 2014;29:613–9.
28. Jimenez-Jimenez FJ, Alonso-Navarro H, Garcia-Martin E, Agundez JAG. Alcohol consumption and risk for Parkinson's disease: a systematic review and meta-analysis. J Neurol. 2019;266:1821–34.
29. Boulos C, Yaghi N, El Hayeck R, Heraoui GN, Fakhoury-Sayegh N. Nutritional risk factors, microbiota and Parkinson's disease: what is the current evidence? Nutrients. 2019;11:1896.
30. Amato A, Liotta R, Mulè F. Effects of menthol on circular smooth muscle of human colon: analysis of the mechanism of action. Eur J Pharmacol. 2014;740:295–301.
31. Micklefield GH, Greving I, May B. Effects of peppermint oil and caraway oil on gastroduodenal motility. Phytother Res. 2000;14:20–3.
32. Cash BD, Epstein MS, Shah SM. A novel delivery system of peppermint oil is an effective therapy for irritable bowel syndrome symptoms. Dig Dis Sci. 2016;61:560–71.
33. López-Liria R, Parra-Egeda J, Vega-Ramírez FA, Aguilar-Parra JM, Trigueros-Ramos R, Morales-Gázquez MJ, Rocamora-Pérez P. Treatment of dysphagia in Parkinson's disease: a systematic review. Int J Environ Res Public Health. 2020;17(11):4104. https://doi.org/10.3390/ijerph17114104. PMID: 32526840; PMCID: PMC7312221.
34. Noelting J, Eaton JE, Choung RS, Zinsmeister AR, Locke GR 3rd, Bharucha AE. The incidence rate and characteristics of clinically diagnosed defecatory disorders in the community. Neurogastroenterol Motil. 2016;28(11):1690–7. https://doi.org/10.1111/nmo.12868. Epub 2016 Jun 2. PMID: 27254309; PMCID: PMC508316.
35. Rao SS, Benninga MA, Bharucha AE, Chiarioni G, Di Lorenzo C, Whitehead WE. ANMS-ESNM position paper and consensus guidelines on biofeedback therapy for anorectal disorders. Neurogastroenterol Motil. 2015;27(5):594–609. https://doi.org/10.1111/nmo.12520. Epub 2015 Apr 1. PMID: 25828100; PMCID: PMC4409469.
36. Rao SS, Rattanakovit K, Patcharatrakul T. Diagnosis and management of chronic constipation in adults. Nat Rev Gastroenterol Hepatol. 2016;13(5):295–305. https://doi.org/10.1038/nrgastro.2016.53. Epub 2016 Apr 1.
37. Wald A, Bharucha AE, Limketkai B, Malcolm A, Remes-Troche JM, Whitehead WE, Zutshi M. ACG clinical guidelines: management of benign anorectal disorders. Am J Gastroenterol. 2021;116(10):1987–2008. https://doi.org/10.14309/ajg.0000000000001507.
38. Woodward S, Norton C, Chiarelli P. Biofeedback for treatment of chronic idiopathic constipation in adults. Cochrane Database Syst Rev. 2014;2014(3):CD008486. https://doi.org/10.1002/14651858.CD008486.pub2. PMID: 24668156; PMCID: PMC10618629.
39. Khera AJ, Chase JW, Salzberg M, Thompson AJV, Kamm MA. Gut-directed pelvic floor behavioral treatment for fecal incontinence and constipation in patients with inflammatory bowel disease. Inflamm Bowel Dis. 2019;25(3):620–6. https://doi.org/10.1093/ibd/izy344.
40. Rao SSC, Quigley EMM, Chey WD, Sharma A, Lembo AJ. Randomized placebo-controlled phase 3 trial of vibrating capsule for chronic constipation. Gastroenterology. 2023;164(7):1202–1210.e6. https://doi.org/10.1053/j.gastro.2023.02.013. Epub 2023 Feb 21.
41. Koo E, Chey WD, Ezell G, et al. Can a defecation posture modifying device correct an abnormal balloon expulsion test in chronically constipated patients? Clin Gastroenterol Hepatol. 2022;21(3):847–8.
42. Ulsh L, Halawi H, Triadafilopoulos G, Gurland B, Nguyen L, Garcia P, Sonu I, Fernandez-Becker N, Becker L, Sheth V, Neshatian L. Use of a footstool improves rectal balloon expulsion in some patients with defecatory disorders. Neurogastroenterol Motil. 2024;36(7):e14781. https://doi.org/10.1111/nmo.14781. Epub 2024 Mar 15.
43. Park WG, Yeh RW, Triadafilopoulos G. Injection therapies for nonvariceal bleeding disorders of the GI tract. Gastrointest Endosc. 2007;66(2):343–54. https://doi.org/10.1016/j.gie.2006.11.019.

44. Yeh RW, Triadafilopoulos G. Injection therapies for nonbleeding disorders of the GI tract. Gastrointest Endosc. 2006;64(3):399–411.; quiz 389–92. https://doi.org/10.1016/j.gie.2006.04.010.
45. Pamphlett R. Early terminal and nodal sprouting of motor axons after botulinum toxin. J Neurol Sci. 1989;92(2–3):181–92. https://doi.org/10.1016/0022-510x(89)90135-4.
46. Kelly JH. Management of upper esophageal sphincter disorders: indications and complications of myotomy. Am J Med. 2000;108(4):43–6.
47. Blitzer A, Brin MF. Use of botulinum toxin for diagnosis and management of cricopharyngeal achalasia. Otolaryngol Head Neck Surg. 1997;116(3):328–30.
48. Shaw GY, Searl JP. Botulinum toxin treatment for cricopharyngeal dysfunction. Dysphagia. 2001;16(3):161–7.
49. Zikos TA, Triadafilopoulos G, Clarke JO. Esophagogastric junction outflow obstruction: current approach to diagnosis and management. Curr Gastroenterol Rep. 2020;22(2):9. https://doi.org/10.1007/s11894-020-0743-0.
50. Shahsavari D, Malik Z, Parkman HP. Management of the patient with esophagogastric junction outflow obstruction. Curr Opin Gastroenterol. 2021;37(4):397–407. https://doi.org/10.1097/MOG.0000000000000747.
51. Miller LS, Szych GA, Kantor SB, Bromer MQ, Knight LC, Maurer AH, Fisher RS, Parkman HP. Treatment of idiopathic gastroparesis with injection of botulinum toxin into the pyloric sphincter muscle. Am J Gastroenterol. 2002;97(7):1653–60. https://doi.org/10.1111/j.1572-0241.2002.05823.x.
52. Triadafilopoulos G, Gandhy R, Barlow C. Pilot cohort study of endoscopic botulinum neurotoxin injection in Parkinson's disease. Parkinsonism Relat Disord. 2017;44:33–7. https://doi.org/10.1016/j.parkreldis.2017.08.020. Epub 2017 Aug 23.
53. Malik Z, Kataria R, Modayil R, Ehrlich AC, Schey R, Parkman HP, Stavropoulos SN. Gastric per oral endoscopic myotomy (G-POEM) for the treatment of refractory gastroparesis: early experience. Dig Dis Sci. 2018;63(9):2405–12. https://doi.org/10.1007/s10620-018-4976-9. Epub 2018 Feb 22.
54. Soliman H, Oiknine E, Cohen-Sors B, Moszkowicz D, Gorbatchef C, Dior M, Nebunu N, Le Gall M, Coffin B, Duboc H. Efficacy and safety of endoscopic pyloric balloon dilation in patients with refractory gastroparesis. Surg Endosc. 2022;36(11):8012–20. https://doi.org/10.1007/s00464-022-09230-w. Epub 2022 Apr 18.
55. Revicki DA, Rentz AM, Dubois D, Kahrilas P, Stanghellini V, Talley NJ, Tack J. Gastroparesis cardinal symptom index (GCSI): development and validation of a patient reported assessment of severity of gastroparesis symptoms. Qual Life Res. 2004;13(4):833–44. https://doi.org/10.1023/B:QURE.0000021689.86296.e4. PMID: 15129893.
56. Chaichanavichkij P, Vollebregt PF, Scott SM, Knowles CH. Botulinum toxin type A for the treatment of dyssynergic defaecation in adults: a systematic review. Color Dis. 2020;22(12):1832–41. https://doi.org/10.1111/codi.15120. Epub 2020 Jun 11.
57. Salvi F, Petrino R, Conroy SP, Liperoti R, Paoletti L, Beccacece A, dell'Aquila G, Fedecostante M, Cherubini A. Constipation: a neglected condition in older emergency department patients. Intern Emerg Med. 2024;19:1977.
58. Metta V, Leta V, Mrudula KR, Prashanth LK, Goyal V, Borgohain R, Chung-Faye G, Chaudhuri KR. Gastrointestinal dysfunction in Parkinson's disease: molecular pathology and implications of gut microbiome, probiotics, and fecal microbiota transplantation. J Neurol. 2022;269(3):1154–63. https://doi.org/10.1007/s00415-021-10567-w. Epub 2021 Apr 21.

8 Gut Emergencies

Acute Dysphagia

Difficulty swallowing (dysphagia) is quite prevalent among patients with PD. Although usually of gradual onset and slowly progressing over months or years, dysphagia may present acutely, typically in the form of complete inability to swallow (aphagia). Foreign body (typically partial dentures) and food bolus (such as dry meat) impaction should be always considered, particularly if the patient suddenly cannot handle their own saliva with ensuing need to constantly spit. Drug-induced esophageal injury and, rarely, levodopa toxicity leading to obstruction need to be considered [1]. Urgent endoscopy should be undertaken particularly if associated with acute onset of painful swallowing (odynophagia).

Colonic Volvulus

Sigmoid volvulus is a common emergency, especially in elderly patients with PD. Patients can variably present with mild abdominal pain and distention to frank peritonitis secondary to colonic perforation. These patients generally need urgent treatment, be it endoscopic decompression of the colon [2] or colonic resection (colectomy) [3]. Volvulus may occur either in the sigmoid colon or the cecum and causes large bowel obstruction. It presents with acute diffuse abdominal pain and distention, nausea, vomiting, and inability to pass stool or gas through the rectum. If untreated, it can progress to bowel gangrene. Sigmoid volvulus affects older patients with PD (mean age > 70), who are debilitated by long-standing constipation and colonic dysmotility. Mortality can be quite high, around 10% for uncomplicated cases and up to 60% if bowel gangrene develops.

The presentation can be acute or subacute, associated with progressively worsening abdominal pain. A plain abdominal radiograph can be obtained first but only establishes the diagnosis in 60% of cases. Often times an abdominal

G. Triadafilopoulos, *The Gut in Parkinson's Disease*,
https://doi.org/10.1007/978-3-031-77171-2_8

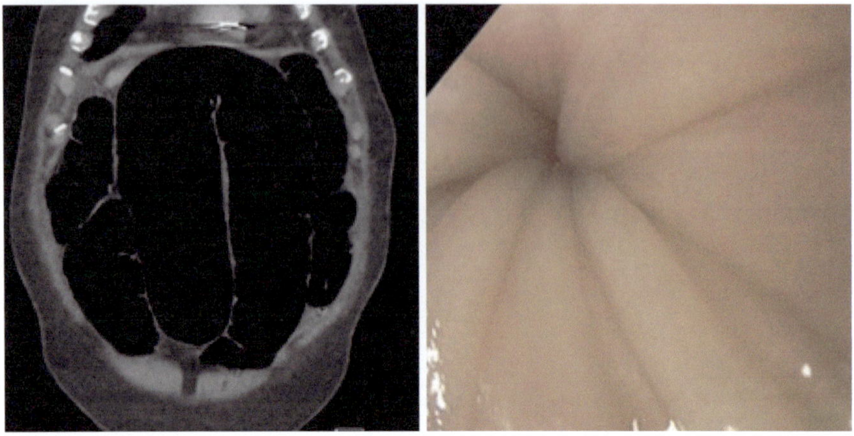

Fig. 8.1 Left: Coronal abdominal CT image of sigmoid volvulus in a PD patient. Right: Colonoscopic view of the neck of the volvulus. Reduction of the volvulus was accomplished upon crossing this point

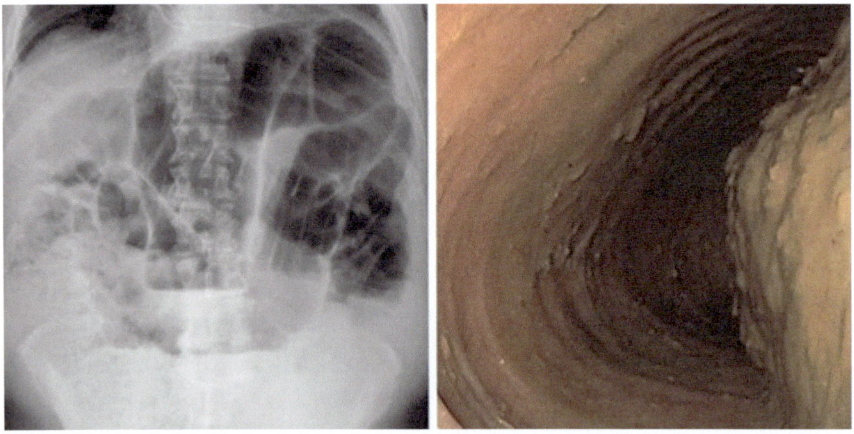

Fig. 8.2 Left: Plain abdominal radiograph of a PD patient presenting with subacute abdominal distention revealing dilated colonic loops. Right: Colonoscopic image of the rectosigmoid of the same patient revealing impacted stool. Disimpaction and extensive colonic lavage was curative

CT scan is needed, which will also rule out other causes of obstruction. Management is either endoscopic reduction or surgical resection. If bowel necrosis is suspected, emergent surgery is mandatory. If bowel necrosis is not suspected, nonsurgical reduction with colonoscopy is preferred, which has been successful in up to 75% of cases. However, recurrence can happen in up to 57% of these patients. Elective resection, after initial endoscopic reduction, may be needed to prevent recurrence (Figs. 8.1 and 8.2).

Fecal Impaction

Prolonged constipation in PD patients can lead to fecal impaction and, less often, stercoral colitis (bowel necrosis caused by pressure ulceration) which can carry an excessively high mortality. A digital rectal examination should be performed on all patients with suspicion of impaction for both diagnosis and management. However, impaction can also occur in the proximal rectum and sigmoid colon so an abdominal radiograph should be obtained if digital rectal examination is negative. Management consists of manual stool disimpaction, liberal enema use, and establishment of an effective long-term maintenance bowel regimen. If manual disimpaction and enemas fail, colonoscopic disimpaction should be pursued.

Ogilvie's Syndrome

Ogilvie's syndrome, or acute intestinal pseudo-obstruction, manifests with symptoms of colonic obstruction, such as bloating and abdominal distension, but in the absence of any mechanical blockage. A number of associated conditions exist, including electrolyte disturbances, infections, major surgeries (particularly orthopedic), and use of narcotic medications. A small study reported an incidence of 2.4% among PD patients. Diagnosis is made by plain abdominal radiograph or abdominal CT, which is necessary to rule out mechanical obstruction. Management is often conservative therapy at first with IV fluids and monitoring for life-threatening complications, such as bowel ischemia and perforation. Intravenous neostigmine, monitored by telemetry and quick access to atropine, is the next option if conservative measures fail. If narcotics are being used, peripheral opioid antagonists, such as intravenous methylnaltrexone, should be considered. If there is still no improvement, colonoscopic decompression should be attempted. Surgical or percutaneous decompression should be reserved only for refractory cases or in emergent situations.

References

1. Abdi S, Masbough F, Nazari M, Abbasinazari M. Drug-induced esophagitis and helpful management for healthcare providers. Gastroenterol Hepatol Bed Bench. 2022;15(3):219–24. https://doi.org/10.22037/ghfbb.v15i3.2591.
2. Aboelezz AF, Othman MO. Endoscopic Management of Colonic Obstruction. Gastrointest Endosc Clin N Am. 2024;34(1):141–53. https://doi.org/10.1016/j.giec.2023.09.011. Epub 2023 Oct 13.
3. Tian BWCA, Vigutto G, Tan E, van Goor H, Bendinelli C, Abu-Zidan F, Ivatury R, Sakakushev B, Di Carlo I, Sganga G, Maier RV, Coimbra R, Leppäniemi A, Litvin A, Damaskos D, Broek RT, Biffl W, Di Saverio S, De Simone B, Ceresoli M, Picetti E, Galante J, Tebala GD, Beka SG, Bonavina L, Cui Y, Khan J, Cicuttin E, Amico F, Kenji I, Hecker A, Ansaloni L, Sartelli M, Moore EE, Kluger Y, Testini M, Weber D, Agnoletti V, Angelis ND, Coccolini F, Sall I, Catena F. WSES consensus guidelines on sigmoid volvulus management. World J Emerg Surg. 2023;18(1):34. https://doi.org/10.1186/s13017-023-00502-x.

Surgery

9

Whenever possible, patients with possible gut PD should be evaluated clinically and by using various relevant structural and functional tests to examine for the underlying presence of focal or diffuse dysfunction predisposing to perioperative adverse effects. Importantly, special guidance needs to be provided by the movement disorder specialist who will need to adjust the medications used for PD during the perioperative period (Fig. 9.1).

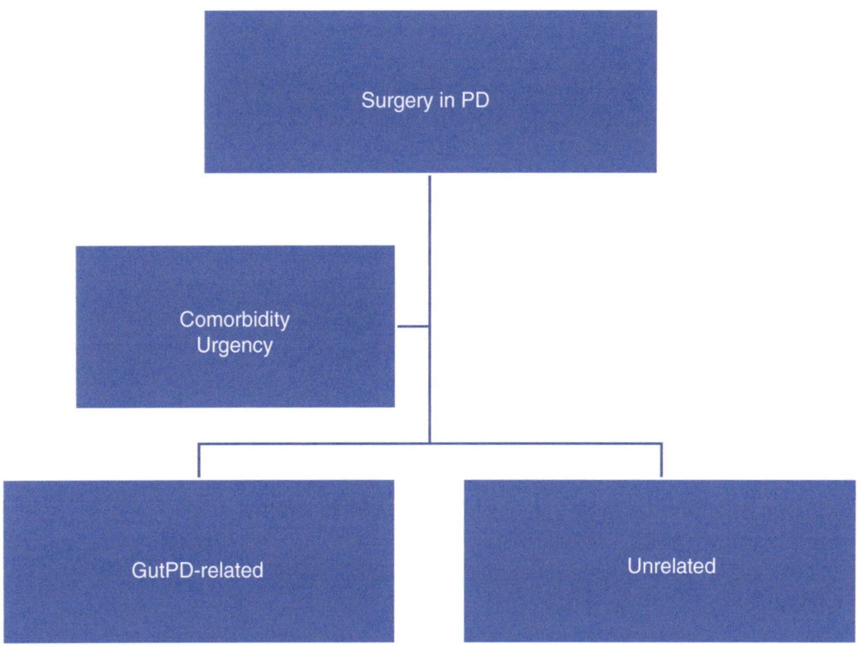

Fig. 9.1 Key considerations in surgery in patients with PD

G. Triadafilopoulos, *The Gut in Parkinson's Disease*,
https://doi.org/10.1007/978-3-031-77171-2_9

There are also several gut PD-related conditions that may require elective or urgent surgery. In all cases, the underlying comorbidity and urgency will need to be considered. Preferably, other medical or endoscopic options need to be explored first, again considering the benefit-to-risk ratio for each patient and what the long-term implications of such surgery might be. Key gut PD-related conditions and their surgical implications are discussed below.

Esophageal Perforation

Esophageal perforation is rare, but potentially fatal, and requires prompt diagnosis and treatment. In patients with gut PD, esophageal perforation may be iatrogenic, typically in the context of food bolus impaction and efforts toward endoscopic disimpaction, or spontaneously resulting from retching (Boerhaave's syndrome) and underlying severe gastroparesis. The clinical presentation depends on the cause and location of the perforation, as well as the length of time since the perforation. Cervical perforation can present with neck pain, erythema, and edema, whereas thoracic perforations can present with chest pain, and gastroesophageal junction perforations can present with either chest or abdominal pain or both. Subcutaneous emphysema and signs of sepsis and multiorgan failure may be seen. A high index of suspicion, a complete history, and physical examination are key. Imaging involves a computed tomography scan of the neck and chest with oral contrast. The gold standard is a fluoroscopic contrast esophagram using gastrografin, followed by thin barium if the diagnosis is not confirmed.

Surgery has traditionally been required for treatment, but endoscopic stenting and endoscopic vacuum therapy are increasingly used therapies that are safe and effective in many patients [1]. Primary surgical repair may be possible but is contraindicated in diffuse mediastinal necrosis, or a large perforation preexisting severe benign esophageal disease, and clinical instability. All patients suspected of esophageal perforation should be fasted and started on appropriate intravenous antimicrobial therapy. For those who are hemodynamically unstable or show signs of clinical deterioration, operative debridement and control of contamination is recommended [2].

Acute Gastric Volvulus

Gastric volvulus is a life-threatening condition, requiring prompt diagnosis and treatment. It is characterized by abnormal rotation of the stomach of more than 180°. As such, it carries the risk of strangulation resulting in necrosis, perforation, hypovolemic shock, and a mortality of 30–50% [3]. The clinical presentation may be acute or chronic and depends upon the speed of onset, the type of volvulus, and the degree of obstruction. Those with the acute form present with pain in the upper abdomen or lower chest associated with severe retching and inability to pass a nasogastric tube.

Acutely, open surgical reduction with or without gastropexy is most frequently performed. However, several other operative approaches may be used, including diaphragmatic hernia repair, gastropexy, partial gastrectomy, or fundo-antral gastro-gastrostomy. However, in the elderly population with PD or those with chronic gastric volvulus, laparoscopic surgery or endoscopic reduction with insertion of a percutaneous gastrostomy tube is used [4].

Severe Gastroparesis

High-Frequency Gastric Electrical Stimulation (GES)

There is very limited data on the use of high-frequency gastric electrical stimulation for patients with PD-related gastroparesis. The mechanism of action of high-frequency (12 cpm) gastric neurostimulation is uncertain and probably multifactorial, since neurostimulation has no effect on basal gastric electrical activity, but it may enhance slow-wave amplitude and propagation velocity [5]. Yet, gastric stimulation increases maximally tolerated volumes of gastric distention, suggesting reduced luminal perception [6]. Response rates to gastric neurostimulation range from 50 to 90%, potentially persisting for up to 10 years [7]. In general, nausea and vomiting are more likely to improve than abdominal pain, fullness, bloating, or acid reflux symptoms [8].

The permanent gastric neurostimulator (Enterra) is implanted through laparoscopy under concomitant endoscopic guidance. The device consists of a pair of leads, a pulse generator, and a programmer. The leads are placed in the muscularis propria of the greater curvature of the stomach, 10 cm proximal to the pylorus, and are connected to a pulse generator which is placed subcutaneously in the left upper quadrant of the abdomen. An external programming device controls the stimulation parameters. Battery life is typically 5–10 years, depending on the energy settings. Endoscopically placed temporary stimulating electrodes have been used to predict response to the permanent device [9].

Pyloromyotomy or pyloroplasty improves patient response better than with GES. There are minor improvements in nausea and abdominal pain with GES as compared to pyloromyotomy or pyloroplasty and lesser antiemetic effects than seen after pyloric surgery or gastrectomy. For GES, vomiting is more likely to improve than abdominal pain [10]. Similarly, clinical responses to GES are less than with gastric peroral endoscopic pyloromyotomy (G-POEM) [11].

Nonocclusive Mesenteric Ischemia (NOMI)

Nonocclusive mesenteric ischemia is an acute and severe pathological condition characterized by signs and symptoms of bowel obstruction and intestinal necrosis, resulting from acute and/or chronic inadequate blood perfusion, in the absence of an organic vascular obstruction detectable by imaging techniques. NOMI is very rare but has a poor prognosis with a 70–90% mortality. Combined treatment with high doses of levodopa and vasodilators, such as PGE (prostaglandin E), may improve prognosis [12].

Severe Refractory Constipation

Subtotal colectomy with ileorectal anastomosis (ileorectostomy) can dramatically improve refractory slow transit constipation in well-studied patients, but it will not be helpful in most patients with gut PD who typically suffer from pelvic floor dyssynergia [13]. Colectomy should be only performed in patients with chronic, severe, and disabling constipation that is unresponsive to medical therapy, who are proven to have slow transit by either ROM or WMC. Ideally, albeit rarely the case, patients should not have pelvic floor dysfunction, based on anorectal manometry, balloon expulsion testing, or defecography. Unfortunately, most patients with gut PD have overlapping features of STC and DD, and a careful assessment is imperative. Further, patients should not have evidence of chronic intestinal pseudo-obstruction (CIPO), as assessed by radiologic (MRE) or manometric studies (GDM), suggesting underlying gastric or small intestinal transit delay, or present with chronic abdominal pain as their prominent symptom, because such patients have poor outcomes. Overall, such carefully selected patients report long-term, very high satisfaction and relief [14]. If colectomy is performed, prolonged postoperative ileus and later episodes of small bowel obstruction may occur and may be challenging to recognize and manage.

Constipated patients with gut PD may be found to have rectocele and/or rectal intussusception on MR defecography. Caution should be given not to rush to operative repair, since surgery may not alleviate their symptoms. Further tests to exclude pelvic floor dysfunction and pelvic floor PT should be done prior to surgery. In women with gut PD and constipation, a useful test is to show improved rectal evacuation upon application of pressure on the posterior wall of the vagina during defecation, before considering a rectocele repair.

Almost 20% of patients with gut PD and chronic constipation may exhibit absence of the recto-anal inhibitory reflex (RAIR) during HRARM. Such lack of relaxation of the internal anal sphincter with balloon rectal distension is suggestive of Hirschsprung's disease (HD), typically a pediatric disease, but false positives can occur, particularly in long-standing cases [15]. The mainstay of treatment for HD is surgery, aiming at resecting the affected segment of the rectum and colon, bringing the normal ganglionic bowel down to a distal rectal anastomosis, close to the anus, and preserving internal anal sphincter function [16]. In patients with gut PD, the diagnosis of adult HD needs to be established prior to surgical decision-making, by demonstrating that ganglion cells are absent in deep rectal biopsies, at times showing hypertrophic nerve fibers, increased acetylcholinesterase activity or staining in the muscularis mucosae, and decreased or absent calretinin-immunoreactive fibers in the lamina propria [17]. A normal rectal biopsy virtually excludes HD, provided that the biopsy samples are obtained from the correct site (within 2 cm from the anorectal junction) and contain muscularis mucosae (typically and safely obtained by snare biopsies after submucosal lifting).

Colonic Volvulus

Colonic volvulus, or the twisting of a segment of colon on its mesentery, is a common surgical emergency, especially in elderly patients and, particularly, those with PD. These patients generally need urgent treatment, either by endoscopic colonic decompression or colectomy. The sigmoid colon is involved in 60–75% of cases [18]. Symptoms include abdominal pain, obstipation, inability to pass gas, and vomiting. In 30–40% of cases, patients report previous episodes of abdominal distention [19].

Physical examination may identify abdominal distension, diminished bowel sounds, and often an empty rectum on digital examination. The absence of peritoneal signs does not indicate the absence of bowel ischemia. Symptoms may vary in duration from a few hours to several days. Patients with PD and cognitive impairment may not be forthcoming or accurate in their symptom description, and the physical examination and testing of electrolytes, blood gases, and lactate levels are crucial [20].

Plain abdominal radiographs are often diagnostic of sigmoid volvulus. Chest radiographs may also detect free air, in cases of perforation. The classic finding is that of a coffee bean, projecting toward the upper abdomen. CT imaging can be used in cases where the diagnosis is in doubt or if ischemia or perforation is suspected. In such cases, flexible endoscopy should be performed as a first line to decompress the sigmoid colon. Such urgent endoscopic detorsion, which is effective in 60–95% of patients, has a 3–4% morbidity and mortality [21]. Successful detorsion implies that the endoscopist must visualize and go past the transition points (typically two points) to fully assess mucosal viability. Afterward, a decompression tube should be left in place to sustain the reduction, allow for continued colonic decompression, and facilitate mechanical bowel preparation [22]. If advanced mucosal ischemia, perforation, and impending perforation of the colon are discovered, an emergency colectomy is warranted.

Even for patients with successful endoscopic decompression, sigmoid colectomy should be offered to prevent recurrences that may occur in 40–70% of cases. The colectomy should be performed as early as possible, even during the index admission. Surgical detorsion, or sigmoidoplasty, should be avoided. Patients who have concomitant megacolon and sigmoid volvulus, frequently seen in patients with long-standing PD, should undergo subtotal colectomy to avoid recurrences in the residual colon [23].

Intestinal Perforation

Elderly patients with PD with long-standing history of constipation may develop stercoral ulcer and colonic perforation with peritonitis. If not suspected in a timely fashion, assessed by abdominal imaging, and attended surgically, it carries high mortality [24].

Acute Colonic Pseudo-obstruction (ACPO)

Acute colonic pseudo-obstruction (ACPO) is a functional dysmotility resulting in colonic distension in the absence of mechanical obstruction that is often referred to as Ogilvie's syndrome [25]. Patients with ACPO present with abdominal distention and abdominal pain, with absence of flatus or stool, along with nausea and vomiting. The abdomen is distended on examination. The diagnosis of ACPO is based on abdominal radiography, or preferably on CT scan with oral, intravenous, and rectal contrast.

Surgery is typically reserved for complications of ACPO. Patients suspected to have colonic ischemia or perforation require operative intervention, usually a laparotomy, or laparoscopy [26]. A right hemicolectomy and end ileostomy with or without mucous fistula is an appropriate procedure to perform on patients who have perforated the cecum in the setting of ACPO. Cecostomy may still have a role in some patients for whom medical and endoscopic decompression fails; the procedure can be performed via endoscopic or percutaneous routes or a combination of both. There is little role for open cecostomy outside resource-limited settings. Complications associated with cecostomy tubes include sepsis, leakage, and catheter dislodgement [27]. Some patients with PD may have recurrent symptoms that can be especially difficult to treat because of an underlying slow transit dysmotility or chronic intestinal pseudo-obstruction.

Fecal Impaction (FI)

Fecal impaction (FI) is defined by a large mass of stool in the rectum and/or colon that cannot be evacuated. FI is more prevalent with increasing age and in patients with gut PD or other neurological disorders. A digital rectal examination should be performed on all patients with suspected fecal impaction for both diagnosis and management. However, impaction can also occur in the proximal rectum and sigmoid colon so an abdominal radiograph should be obtained if digital rectal examination is negative. Management consists of manual disimpaction, liberal enema use, and establishment of an effective long-term maintenance bowel regimen. Abdominal surgery, both for bowel resections for ischemia/necrosis of the small bowel and colon, is associated with very high mortality [28].

Prolonged and inadequately treated constipation in patients with gut PD may lead not only to fecal impaction but, rarely, stercoral *colitis*, an uncommon but potentially life-threatening condition that can present acutely, characterized by impacted fecal material causing increased colonic intraluminal pressure and inflammation. If not recognized and treated, stercoral colitis may lead to dangerous complications such as ischemic necrosis, perforation, and/or sepsis, with a mortality rate up to 63% [29].

Deep Brain Stimulation (DBS)

Deep brain stimulation (DBS) therapy involves the intracranial placement of thin leads (on one side or both sides of the brain which are connected through extension leads to a neurostimulator that is placed under the skin beneath the clavicle) [30]. Electrical stimulation can be adjusted and tailored to target an individual patient's symptoms over time. The mechanism of action is not fully known. Most patients undergoing DBS therapy have motor fluctuations that are poorly controlled by medication. Determination of candidacy for DBS, implantation of the system, and ongoing patient care and device management are typically performed at specialized centers for PD.

After device parameter optimization, DBS improves the quality of life and alleviates motor fluctuations with an average increase in "on" time, of 3–4 h a day, and it also allows for a 50% reduction in medication. Given the low procedural risks (i.e., stroke and infection) with modern procedures and its effectiveness, DBS is approved for use when motor fluctuations begin [31]. The motor benefit may persist for up to 15 years.

Some DBS systems include image-based software that allow visualization of the leads relative to the targeted brain structure and simulation of the volume of tissue activation to facilitate programming. Some sensing systems can also allow for adaptive stimulation, based on neuronal feedback signal. Unfortunately, the non-motor symptoms of PD (i.e., cognition, mood, and autonomic dysfunction) do not improve with the use of DBS.

References

1. Barakat MT, Girotra M, Banerjee S. (Re)building the wall: recurrent Boerhaave syndrome managed by over-the-scope clip and covered metallic stent placement. Dig Dis Sci. 2018;63(5):1139–42. https://doi.org/10.1007/s10620-017-4756-y. Epub 2017 Sep 25.
2. Fairbairn K, Worrell SG. Esophageal perforation: is surgery still necessary? Thorac Surg Clin. 2023;33(2):117–23. https://doi.org/10.1016/j.thorsurg.2023.01.005. Epub 2023 Feb 26.
3. Chau B, Dufel S. Gastric volvulus. Emerg Med J. 2007;24(6):446–7. https://doi.org/10.1136/emj.2006.041947.
4. Eckhauser ML, Ferron JP. The use of dual percutaneous endoscopic gastrostomy (DPEG) in the management of chronic intermittent gastric volvulus. Gastrointest Endosc. 1985;31(5):340–2. https://doi.org/10.1016/s0016-5107(85)72221-3.
5. Lin Z, Forster J, Sarosiek I, McCallum RW. Effect of high-frequency gastric electrical stimulation on gastric myoelectric activity in gastroparetic patients. Neurogastroenterol Motil. 2004;16(2):205–12. https://doi.org/10.1111/j.1365-2982.2004.00503.x.
6. Gourcerol G, Ouelaa W, Huet E, Leroi AM, Ducrotte P. Gastric electrical stimulation increases the discomfort threshold to gastric distension. Eur J Gastroenterol Hepatol. 2013;25(2):213–7. https://doi.org/10.1097/MEG.0b013e32835a7f04.
7. McCallum RW, Lin Z, Forster J, Roeser K, Hou Q, Sarosiek I. Gastric electrical stimulation improves outcomes of patients with gastroparesis for up to 10 years. Clin Gastroenterol Hepatol. 2011;9(4):314–319.e1. https://doi.org/10.1016/j.cgh.2010.12.013. Epub 2010 Dec 23.

8. Maranki JL, Lytes V, Meilahn JE, Harbison S, Friedenberg FK, Fisher RS, Parkman HP. Predictive factors for clinical improvement with Enterra gastric electric stimulation treatment for refractory gastroparesis. Dig Dis Sci. 2008;53(8):2072–8. https://doi.org/10.1007/s10620-007-0124-7. Epub 2007 Dec 14.
9. Abell TL, Johnson WD, Kedar A, Runnels JM, Thompson J, Weeks ES, Minocha A, Griswold ME. A double-masked, randomized, placebo-controlled trial of temporary endoscopic mucosal gastric electrical stimulation for gastroparesis. Gastrointest Endosc. 2011;74(3):496–503.e3. https://doi.org/10.1016/j.gie.2011.05.022.
10. Zoll B, Zhao H, Edwards MA, Petrov R, Schey R, Parkman HP. Outcomes of surgical intervention for refractory gastroparesis: a systematic review. J Surg Res. 2018;231:263–9. https://doi.org/10.1016/j.jss.2018.05.061. Epub 2018 Jun 23.
11. Shen S, Luo H, Vachaparambil C, Mekaroonkamol P, Abdelfatah MM, Xu G, Chen H, Xia L, Shi H, Keilin S, Willingham F, Christie J, Lin E, Cai Q. Gastric peroral endoscopic pyloromyotomy versus gastric electrical stimulation in the treatment of refractory gastroparesis: a propensity score-matched analysis of long-term outcomes. Endoscopy. 2020;52(5):349–58. https://doi.org/10.1055/a-1111-8566. Epub 2020 Feb 21. Erratum in: Endoscopy. 2020 May;52(5):C5.
12. Salamone G, Raspanti C, Licari L, Falco N, Rotolo G, Augello G, Melfa G, Gulotta G. Non-occlusive mesenteric ischemia (NOMI) in Parkinson's disease: case report. G Chir. 2017;38(2):71–6. https://doi.org/10.11138/gchir/2017.38.2.071.
13. Nyam DC, Pemberton JH, Ilstrup DM, Rath DM. Long-term results of surgery for chronic constipation. Dis Colon Rectum. 1997;40(3):273–9. https://doi.org/10.1007/BF02050415. Erratum in: Dis Colon Rectum 1997 May;40(5):529.
14. Pikarsky AJ, Singh JJ, Weiss EG, Nogueras JJ, Wexner SD. Long-term follow-up of patients undergoing colectomy for colonic inertia. Dis Colon Rectum. 2001;44(2):179–83. https://doi.org/10.1007/BF02234290.
15. Su A, Gandhy R, Barlow C, Triadafilopoulos G. Utility of high-resolution anorectal manometry and wireless motility capsule in the evaluation of patients with Parkinson's disease and chronic constipation. BMJ Open Gastroenterol. 2016;3(1):e000118. https://doi.org/10.1136/bmjgast-2016-000118.
16. Mao YZ, Tang ST, Li S. Duhamel operation vs. transanal endorectal pull-through procedure for Hirschsprung disease: a systematic review and meta-analysis. J Pediatr Surg. 2018;53(9):1710–5. https://doi.org/10.1016/j.jpedsurg.2017.10.047. Epub 2017 Oct 16.
17. Lake BD, Puri P, Nixon HH, Claireaux AE. Hirschsprung's disease: an appraisal of histochemically demonstrated acetylcholinesterase activity in suction rectal biopsy specimens as an aid to diagnosis. Arch Pathol Lab Med. 1978;102(5):244–7.
18. Perrot L, Fohlen A, Alves A, Lubrano J. Management of the colonic volvulus in 2016. J Visc Surg. 2016;153(3):183–92. https://doi.org/10.1016/j.jviscsurg.2016.03.006. Epub 2016 Apr 28.
19. Gingold D, Murrell Z. Management of colonic volvulus. Clin Colon Rectal Surg. 2012;25(4):236–44. https://doi.org/10.1055/s-0032-1329535.
20. Bruzzi M, Lefèvre JH, Desaint B, Nion-Larmurier I, Bennis M, Chafai N, Tiret E, Parc Y. Management of acute sigmoid volvulus: short- and long-term results. Color Dis. 2015;17(10):922–8. https://doi.org/10.1111/codi.12959.
21. Renzulli P, Maurer CA, Netzer P, Büchler MW. Preoperative colonoscopic derotation is beneficial in acute colonic volvulus. Dig Surg. 2002;19(3):223–9. https://doi.org/10.1159/000064217.
22. Oren D, Atamanalp SS, Aydinli B, Yildirgan MI, Başoğlu M, Polat KY, Onbaş O. An algorithm for the management of sigmoid colon volvulus and the safety of primary resection: experience with 827 cases. Dis Colon Rectum. 2007;50(4):489–97. https://doi.org/10.1007/s10350-006-0821-x.
23. Tian BWCA, Vigutto G, Tan E, van Goor H, Bendinelli C, Abu-Zidan F, Ivatury R, Sakakushev B, Di Carlo I, Sganga G, Maier RV, Coimbra R, Leppäniemi A, Litvin A, Damaskos D, Broek RT, Biffl W, Di Saverio S, De Simone B, Ceresoli M, Picetti E, Galante J, Tebala GD, Beka SG, Bonavina L, Cui Y, Khan J, Cicuttin E, Amico F, Kenji I, Hecker A, Ansaloni L, Sartelli

M, Moore EE, Kluger Y, Testini M, Weber D, Agnoletti V, Angelis ND, Coccolini F, Sall I, Catena F. WSES consensus guidelines on sigmoid volvulus management. World J Emerg Surg. 2023;18(1):34. https://doi.org/10.1186/s13017-023-00502-x.
24. Tateno F, Sakakibara R, Aiba Y, Tsuyusaki Y, Kishi M, Tateno H, Ogata T. Stercoral ulcer and colonic perforation in an individual with Parkinson's disease with constipation. J Am Geriatr Soc. 2016;64(10):e118–20. https://doi.org/10.1111/jgs.14357.
25. Arthur T, Burgess A. Acute colonic pseudo-obstruction. Clin Colon Rectal Surg. 2022;35(3):221–6. https://doi.org/10.1055/s-0041-1740044.
26. Duh QY, Way LW. Diagnostic laparoscopy and laparoscopic cecostomy for colonic pseudo-obstruction. Dis Colon Rectum. 1993;36(1):65–70. https://doi.org/10.1007/BF02050304.
27. Vogel JD, Feingold DL, Stewart DB, Turner JS, Boutros M, Chun J, Steele SR. Clinical practice guidelines for colon volvulus and acute colonic pseudo-obstruction. Dis Colon Rectum. 2016;59(7):589–600. https://doi.org/10.1097/DCR.0000000000000602.
28. Sommers T, Petersen T, Singh P, Rangan V, Hirsch W, Katon J, Ballou S, Cheng V, Friedlander D, Nee J, Lembo A, Iturrino J. Significant morbidity and mortality associated with fecal impaction in patients who present to the emergency department. Dig Dis Sci. 2019;64(5):1320–7. https://doi.org/10.1007/s10620-018-5394-8. Epub 2018 Dec 8.
29. Bae E, Tran J, Shah K. Stercoral colitis in the emergency department: a review of the literature. Int J Emerg Med. 2024;17(1):3. https://doi.org/10.1186/s12245-023-00578-x.
30. Tanner CM, Ostrem JL. Parkinson's disease. N Engl J Med. 2024;391(5):442–52.
31. Schuepbach WM, Rau J, Knudsen K, Volkmann J, Krack P, Timmermann L, Hälbig TD, Hesekamp H, Navarro SM, Meier N, Falk D, Mehdorn M, Paschen S, Maarouf M, Barbe MT, Fink GR, Kupsch A, Gruber D, Schneider GH, Seigneuret E, Kistner A, Chaynes P, Ory-Magne F, Brefel Courbon C, Vesper J, Schnitzler A, Wojtecki L, Houeto JL, Bataille B, Maltête D, Damier P, Raoul S, Sixel-Doering F, Hellwig D, Gharabaghi A, Krüger R, Pinsker MO, Amtage F, Régis JM, Witjas T, Thobois S, Mertens P, Kloss M, Hartmann A, Oertel WH, Post B, Speelman H, Agid Y, Schade-Brittinger C, Deuschl G, EARLYSTIM Study Group. Neurostimulation for Parkinson's disease with early motor complications. N Engl J Med. 2013;368(7):610–22.

Evolution

10

Factors Affecting Gut PD Severity

Patients with PD frequently require changes in the timing of administration of their levodopa–carbidopa, altering of the dose, and switching to alternatives or adjunctive agents or even other interventions (i.e., intrajejunal levodopa administration) for their motor disease [1]. The management of gut PD remains challenging to neurologists (particularly when symptoms are a result of anti-Parkinson's medications) and gastroenterologists. Unfortunately, there are limited data on the options available, if prior therapies fail. Clinicians need to use their best judgment based on a thorough assessment of the patients' history and objectively demonstrable pathophysiological abnormalities. Since they address various targets, combination therapies, including acid suppressants, promotility agents, and laxatives, may play a role in restoring quality of life and minimizing symptoms. Faced with refractory gut PD, the physician needs to consider the effects of disease on symptoms and quality of life, the burden it imposes to the entire digestive system, both structurally and functionally, and the overall course of the disease and its complications [2].

The flow diagram below highlights such considerations and their clinical impact if left unattended (Fig. 10.1).

Depending on the disease stage, several questions need to be addressed and are highlighted below (Fig. 10.2).

Simultaneously, patients with gut PD would benefit physically and emotionally by coming to terms with an understanding of the impact of gut PD. Based on this, patients can form a multidisciplinary clinical team to manage them long-term, maintaining the locus of control of their decision-making, based on ongoing education about their disease and its associated comorbidities (Fig. 10.3).

The overall objectives in gut PD (as shown below) start with elaborate characterization of the patient's disease presentation, followed by a realistic discussion and implementation of treatment goals and expectations by both the movement disorder

G. Triadafilopoulos, *The Gut in Parkinson's Disease*,
https://doi.org/10.1007/978-3-031-77171-2_10

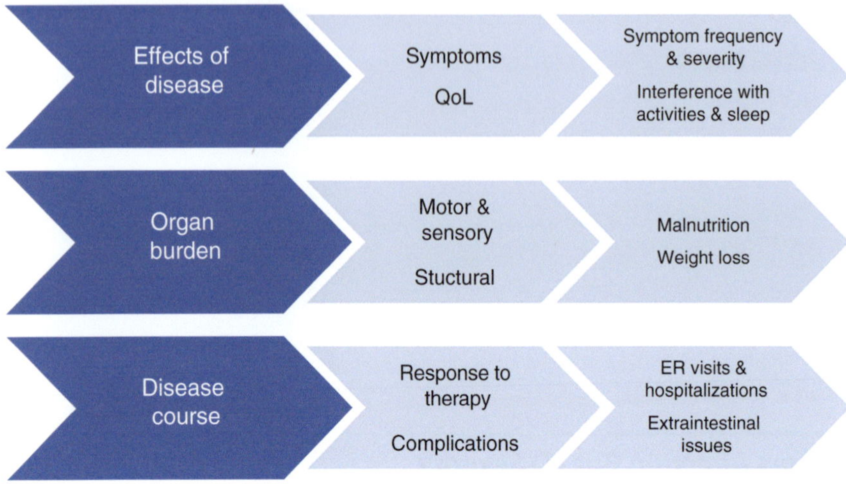

Fig. 10.1 Clinical considerations during the evolution of Gut PD

Early phase	Diagnosis	Initial treatment	Disease fluctuations	End stage
When did gut symptoms start?	What does the gut disease look like?	When is it time for medical therapy?	Can fluctuations be prevented?	What is the end like?
Will other (new) gut symptoms develop?	What causes gut symptoms?	What treatments are there for the gut symptoms?	How do fluctuations get treated?	Is there anything else that can be done?
Can symptoms be prevented?	Are family members at risk?	Are there any non-drug therapies?	Have all options been explored?	Can cachexia be avoided?
Why? Does it run in the family? Toxins?	What is expected to happen?	Is there a need to see a gastroenterologist?	Does diet play a role?	Need for feeding tubes or IVs?

Fig. 10.2 Frequently asked questions (FAQs) during the evolving phases of gut involvement in PD

specialist and gastroenterologist. It is essential to monitor—at least yearly—disease evolution by standardized clinical and multimodality assessment to identify changes, if any, and their pace of evolution. Based on such assessment, relevant modifications may be undertaken (Fig. 10.4).

With the increasing collaboration among neurologists, gastroenterologists, and other related members of the clinical team managing the patient with gut PD, we are slowly moving away from the micromanagement of patients based on their symptoms alone, and we are increasingly incorporating the disease context, severity, and overall impact, not only on the patient's quality of life, but also other important outcomes, such as development of medical or surgical complications, progressive immobility, or even death (Fig. 10.5).

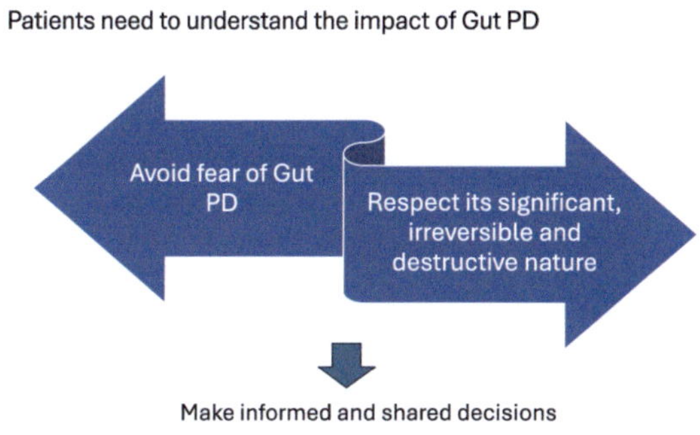

Fig. 10.3 Importance of patients' understanding of the clinical impact of PD

Fig. 10.4 Stepwise approach to clinical objectives in Gut PD management

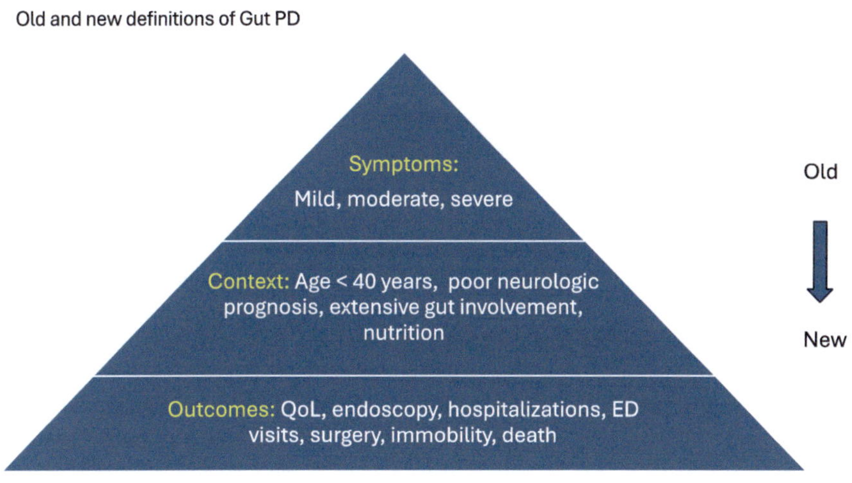

Fig. 10.5 Changing outcomes' definitions in Gut PD over time

Along the disease trajectory of gut PD, it is important to distinguish disease activity from severity. This distinction is essential since it not only provides guidance to clinicians on the overall strategic management of the disease but also facilitates the patient's understanding of the progress made overtime while it identifies potential areas to be pursued (Fig. 10.6).

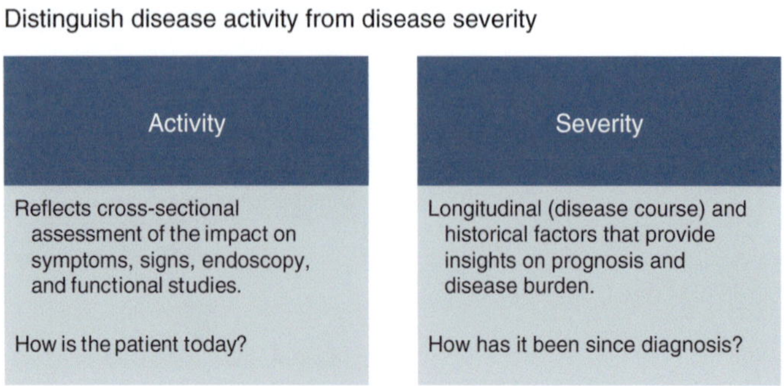

Fig. 10.6 Important questions on Gut PD disease activity and severity

Gut-Related Factors in PD Risk and Progression

There is little evidence supporting a relationship between gut microbiome changes and PD risk and disease phenotype. Constipation, a prodromal symptom of PD, predicts a worse disease evolution. One study in 465 community-based patients with PD, who were monitored for up to 8 years, found a relationship between the presence and severity of constipation at disease onset and faster progression to dementia [3].

Similarly, 423 patients with newly diagnosed PD who were monitored for up to 5 years had higher composite scores for gastrointestinal symptoms (infrequency of bowel movements, hard stools, and fecal incontinence) associated with worse cognitive performance and dementia [4].

Weight loss in PD has been linked to increased risks of malnutrition, fractures, pressure ulcers, and death [5].

Symptoms of irritable bowel syndrome (IBS), such as abdominal pain and altered bowel habits, are common in gut PD and are associated with changes in the fecal microbiota (dysbiosis), increased gut permeability, and low-grade mucosal inflammation, all potentially making the gastrointestinal tract more vulnerable to PD [6]. *H. pylori* infection doubles the risk of developing PD [7]. In contrast, eradication of *H. pylori* is associated with a 45% increased risk of developing PD 5 or more years later [8].

Both *H. pylori* infection and SIBO have been associated with worse motor severity scores and worse motor response complications (delayed and shorter "on" times) [9]. Both *H. pylori* infection and SIBO might contribute to motor dysfunction in PD by reducing levodopa bioavailability or by inducing a chronic inflammatory state that exacerbates the neurodegenerative process, microglial activation, neuronal injury, and death [10]. On the other hand, gut dysmotility and associated autonomic dysfunction could be a manifestation of severe PD and could contribute to SIBO [11].

Diet plays a key role in microbial composition and function. In turn, gut microbes and their metabolic products have important effects on human health [12]. Although each of the main macronutrients and micronutrients may modify the gut microbiome, indigestible carbohydrates (i.e., fiber) which provide energy for microbes in the colon have a particularly prominent role in diet–microbiome interactions [13]. However, many studies have shown conflicting associations between PD risk, macronutrients, micronutrients, specific foods, and dietary patterns.

Obesity

Obesity, a global epidemic, affecting roughly 30% of the world's population and predicted to rise, results from genetic, behavioral, societal, and environmental factors, leading to excessive fat accumulation, due to insufficient energy expenditure, and it can lead to metabolic and cardiovascular diseases, through its effects on fat, skeletal muscle, and the liver. With the progression of socioeconomic status and alterations in lifestyle, an increasing number of global populations are at risk for increasing body mass index (BMI), a crucial obesity indicator, but there is no causal relationship between BMI and PD [14]. The adipose tissue is a complex organ involved in hormone regulation and modulation of metabolism, inflammation, and homeostasis. In turn, obesity is associated with a low-grade inflammatory state and has been linked to neurodegenerative diseases, such as PD. Mechanistically, reduced adipose expandability leads to hypertrophic adipocytes, triggering inflammation, insulin and leptin resistance, blood–brain barrier disruption, altered brain metabolism, neuronal inflammation, brain atrophy, and cognitive decline [15]. Recent evidence links obesity to diminished brain health and neurodegenerative diseases, such as Alzheimer's disease (AD) and Parkinson's disease (PD), both associated with insulin resistance, but the causative mechanism(s) are unclear [16].

Nutrition and Other Modifiable Risk Factors

Several dietary patterns and nutritional supplements have been linked to the development, progression, and symptomatic treatment of PD, but evidence on effectiveness remains inconclusive. Dietary interventions could, analogous to exercise, potentially have a beneficial effect on disease symptoms as well as on the progression of the disease [17]. Modifiable risk factors for Parkinson's disease (PD) are poorly known. To evaluate independent associations of different nutritional components, physical activity, and sedentary behavior and metabolic factors with the risk of PD, a population-based prospective cohort study on 502,017 men and women who were free from PD at baseline was conducted over a median follow-up of 12.8 years. Vigorous physical activity, reduced sitting time, and good sleep quality together with small coffee intake and vitamin D supplementation were found to be potentially neuroprotective lifestyle interventions for the prevention of PD. Small amounts of coffee, red meat, and present smoking were also associated with a lower risk of PD, whereas alcohol intake was associated with slightly higher PD risk [18].

Nutritional Management

The present paradigm for the multidisciplinary management of PD does not include regular nutritional assessment despite research showing that 90% of people living with the disease lack access to basic dietetic services. Since many non-motor symptoms such as dysphagia, constipation, orthostatic hypotension, and weight loss and sarcopenia can be improved through dietary intervention, dietitians are a critical missing piece of the PD management [19].

Integrative Therapies for PD

Imaging studies have shown that aerobic exercise changes brain structure and function, while strength training improves posture and balance. Loneliness is associated with worsening PD severity, but social prescribing and cognitive behavioral therapy may effectively foster connections. Ayurvedic and traditional Chinese medicine practices including yoga, meditation, tai chi, and acupuncture may help improve mobility, mood, sleep, and quality of life. Art therapy enhances visuospatial skills, whereas music and dance therapy can alleviate freezing of gait. Several studies demonstrate successful use of these integrative strategies virtually, thereby improving patient accessibility and participation. PD management has broadened to include integrative approaches combining conventional and complementary therapies. Potential benefits of movement, nutrition, sleep, socialization, and mind–body practices have been confirmed with several recent randomized controlled trials [20].

References

1. Jost WH. A novel treatment option for intrajejunal levodopa administration. Expert Rev Neurother. 2023;23(1):9–13. https://doi.org/10.1080/14737175.2023.2176222. Epub 2023 Feb 10.
2. Warnecke T, Schäfer KH, Claus I, Del Tredici K, Jost WH. Gastrointestinal involvement in Parkinson's disease: pathophysiology, diagnosis, and management. NPJ Parkinsons Dis. 2022;8(1):31. https://doi.org/10.1038/s41531-022-00295-x.
3. Camacho M, Macleod AD, Maple-Grødem J, Evans JR, Breen DP, Cummins G, Wijeyekoon RS, Greenland JC, Alves G, Tysnes OB, Lawson RA, Barker RA, Williams-Gray CH. Early constipation predicts faster dementia onset in Parkinson's disease. NPJ Parkinsons Dis. 2021;7(1):45. https://doi.org/10.1038/s41531-021-00191-w.
4. Jones JD, Rahmani E, Garcia E, Jacobs JP. Gastrointestinal symptoms are predictive of trajectories of cognitive functioning in de novo Parkinson's disease. Parkinsonism Relat Disord. 2020;72:7–12. https://doi.org/10.1016/j.parkreldis.2020.01.009. Epub 2020 Jan 25.
5. Tan AH, Hew YC, Lim SY, Ramli NM, Kamaruzzaman SB, Tan MP, Grossmann M, Ang BH, Tan JY, Manap MAAA, Tay TK, Tan SL, New RP, Fadzli F, Yee EJ, Moy FM, Mahadeva S, Lang AE. Altered body composition, sarcopenia, frailty, and their clinico-biological correlates, in Parkinson's disease. Parkinsonism Relat Disord. 2018;56:58–64. https://doi.org/10.1016/j.parkreldis.2018.06.020. Epub 2018 Jun 13.

6. Pimentel M, Saad RJ, Long MD, Rao SSC. ACG clinical guideline: small intestinal bacterial overgrowth. Am J Gastroenterol. 2020;115(2):165–78. https://doi.org/10.14309/ajg.0000000000000501.
7. Huang HK, Wang JH, Lei WY, Chen CL, Chang CY, Liou LS. Helicobacter pylori infection is associated with an increased risk of Parkinson's disease: a population-based retrospective cohort study. Parkinsonism Relat Disord. 2018;47:26–31. https://doi.org/10.1016/j.parkreldis.2017.11.331. Epub 2017 Nov 13.
8. Nielsen HH, Qiu J, Friis S, Wermuth L, Ritz B. Treatment for helicobacter pylori infection and risk of Parkinson's disease in Denmark. Eur J Neurol. 2012;19(6):864–9. https://doi.org/10.1111/j.1468-1331.2011.03643.x. Epub 2012 Jan 17.
9. Lee WY, Yoon WT, Shin HY, Jeon SH, Rhee PL. Helicobacter pylori infection and motor fluctuations in patients with Parkinson's disease. Mov Disord. 2008;23(12):1696–700. https://doi.org/10.1002/mds.22190.
10. McGee DJ, Lu XH, Disbrow EA. Stomaching the possibility of a pathogenic role for helicobacter pylori in Parkinson's disease. J Parkinsons Dis. 2018;8(3):367–74. https://doi.org/10.3233/JPD-181327.
11. De Pablo-Fernandez E, Tur C, Revesz T, Lees AJ, Holton JL, Warner TT. Association of autonomic dysfunction with disease progression and survival in Parkinson disease. JAMA Neurol. 2017;74(8):970–6. https://doi.org/10.1001/jamaneurol.2017.1125.
12. Gentile CL, Weir TL. The gut microbiota at the intersection of diet and human health. Science. 2018;362(6416):776–80. https://doi.org/10.1126/science.aau5812.
13. Zmora N, Suez J, Elinav E. You are what you eat: diet, health and the gut microbiota. Nat Rev Gastroenterol Hepatol. 2019;16(1):35–56. https://doi.org/10.1038/s41575-018-0061-2.
14. Wang X, Chen H, Chang Z, Zhang J, Xie D. Genetic causal role of body mass index in multiple neurological diseases. Sci Rep. 2024;14(1):7256. https://doi.org/10.1038/s41598-024-57260-2.
15. Neto A, Fernandes A, Barateiro A. The complex relationship between obesity and neurodegenerative diseases: an updated review. Front Cell Neurosci. 2023;17:1294420. https://doi.org/10.3389/fncel.2023.1294420.
16. Kueck PJ, Morris JK, Stanford JA. Current perspectives: obesity and neurodegeneration—links and risks. Degener Neurol Neuromuscul Dis. 2023;13:111–29. https://doi.org/10.2147/DNND.S388579.
17. van der Berg I, Schootemeijer S, Overbeek K, Bloem BR, de Vries NM. Dietary interventions in Parkinson's disease. J Parkinsons Dis. 2024;14(1):1–16. https://doi.org/10.3233/JPD-230366.
18. Veronese N, Nova A, Fazia T, Riggi E, Yang L, Piccio L, Huang BH, Ahmadi M, Barbagallo M, Notarnicola M, Giannelli G, De Pergola G, Stamatakis E, Cereda E, Bernardinelli L, Fontana L. Contribution of nutritional, lifestyle, and metabolic risk factors to Parkinson's disease. Mov Disord. 2024;39(7):1203–12. https://doi.org/10.1002/mds.29778. Epub 2024 Mar 26.
19. Flanagan R, Rusch C, Lithander FE, Subramanian I. The missing piece of the puzzle—the key role of the dietitian in the management of Parkinson's disease. Parkinsonism Relat Disord. 2024;121:106021. https://doi.org/10.1016/j.parkreldis.2024.106021. Epub 2024 Feb 1.
20. Kola S, Subramanian I. Updates in Parkinson's disease integrative therapies: an evidence-based review. Curr Neurol Neurosci Rep. 2023;23(11):717–26. https://doi.org/10.1007/s11910-023-01312-z. Epub 2023 Nov 3.

Care Coordination in Gut PD

11

Patient–Physician Interactions in Gut PD

Since gut PD is multidimensional, with both structural and functional elements (both gastrointestinal and neurological) that frequently interact and complicate both diagnosis and management, it is important for the physician to be fully informed about the severity, acuity, and, to a degree possible, disease prognosis before spending time informing the patient and their family of their findings and outlining a therapeutic plan. Table 11.1 highlights the essential elements of the first interaction between the gastroenterologist and the patient with gut PD. During this pivotal visit, there is no room for uncertainty; it is a time for building trust in the therapeutic relationship which will only remain effective in the long term, if it is based on transparency and perpetual open-mindedness from either side.

As with any chronic illness, effective and encouraging patient–physician interaction remains paramount for all patients with gut PD. Such interaction needs to come from multiple directions (i.e., neurology, gastroenterology) emphasizing the positives while acknowledging any challenges. The following table provides a list of issues in gut PD that require regular consideration when managing such patients who, during their illness, not only may experience a physical but also a cognitive decline. During follow-up, it is often useful to first privately interact with the escorting family member or caretaker to be kept abreast of recent changes in the patient's clinical and mental or psychiatric performance. This information facilitates the subsequent interview and decision-making that ultimately will remain patient-centered and direct (Fig. 11.1).

It is often useful to consider the patients' perspectives and concerns about their disease prior to engaging into a dialogue about diagnostic investigations, therapeutic options, and other management decisions. Table 11.2 shows a list of issues of concern for most patients with gut PD. Although at first glance this list may seem overwhelming, it is time-saving long term since many of these issues can be addressed preemptively. This approach also helps patients build trust toward their gastroenterologist and the other members of their clinical team.

G. Triadafilopoulos, *The Gut in Parkinson's Disease*,
https://doi.org/10.1007/978-3-031-77171-2_11

Well informed based upon the original concepts of the Kübler-Ross model [1] of the five stages of grief, physicians caring for patients with gut PD should keep in mind the various stages of dealing with their illness (Table 11.3).

Table 11.1 Points to be considered when breaking the news about gut PD

Prepare fully; checking notes, test results, and time arranged for follow-up investigations or planned admission.
Break bad news in a suitable room, perhaps with a nurse present. Ensure privacy with no interruptions.
Allow adequate time.
Make sure patients and relatives know who you are and the purpose of the consultation.
Remember that what is hopeful news for you might still be bad news for the patient, and vice versa.
Do not assume that someone else has handled important parts of the information giving.
Do not be falsely reassuring; provide as much positive practical support and information as possible.
Ensure that the patient has enough time to let news sink in - check understanding before they leave.
Offer follow-up appointment, number to call, or addresses or other helpful agencies.
Ensure that they understand that you are working together with other team members in PD management.

Issues to be addressed	Patient-physician exchange examples
Assess patient's awareness of the problem.	Tell me what worries you the most.
Address understanding of prognosis.	We will try to minimize any functional decline.
Clarification of natural history.	We have several ways to control the disease.
Explore patient's emotions towards gut issues.	I can imagine that this is hard for you and family.
Engage family and caretakers in disease understanding.	We should engage a trusted partner in our discussions.
Examine what matters the most.	We can prioritize the control of your constipation.

Fig. 11.1 Issues to be addressed with patients in the context of a new diagnosis of Gut PD

Table 11.2 Key issues that concern patients with gut PD

Fear of incontinence on public transport, shopping, vacation, and work.
Fear of hospitals and treatments.
Effects on employment including limitations on types of job and promotion prospects.
Effects on holiday, medical and life insurance.
Offer follow-up appointment, number to call, or addresses or other helpful agencies.
Ensure that they understand that you are working together with other team members in PD management.

Table 11.3 Stages of dealing with gut PD

Initial **shock and denial** of the diagnosis; **anger** for getting the disease
Realization of possible morbidity & mortality
Loss of hope and dreams of future
Anxiety and uncertainty of prognosis and evolution
Need for daily medications, their effect(s), or treatment failures
Fear of possible need for other treatments (endoscopic & surgical)
Depression for living with chronic disease and absence of cure
Limited lifestyle because of illness
Professional & social isolation, or rejection
Guilt about being a burden to partner or family

Optimizing Patient-Centered Care in Gut PD

The clinical presentation of gut PD is highly heterogeneous and ranges from silent functional disturbance of propulsion, asymptomatic chronic infection by *H. pylori*, to dysphagia, nausea, or chronic constipation. This multitude of covert

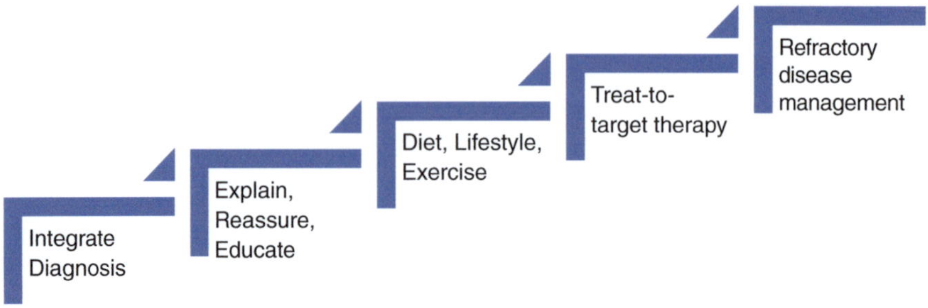

Fig. 11.2 Important elements to optimize a patient-centered, individualized approach to Gut PD

abnormalities and symptoms—that vary in sensitivity and specificity as well as frequency and severity—matched with their underlying multifactorial nature, poses significant challenges in long-term management and prevention of complications. Therefore, to optimize a patient-centered, individualized approach, clinicians need to consider several important elements (Fig. 11.2).

Strategic Management in Gut PD

Strategic management of gut PD should continuously strive to maintain the best possible gut-related quality of life (QoL) in the context of a constantly, but slowly, changing neurologic disease. Together, healthcare providers, patients, supporting family, and caretakers should develop a successful road map by adopting a management strategy that fits their mutually agreed upon short- and long-term health objectives, based on available resources, given the physical and emotional impact and demands of a chronic, degenerative illness. Table outlines the key advantages of this approach (Table 11.4).

Clear sense of direction: Setting specific, achievable directives helps define the patient's perspective of their gut symptoms and defines a gut-specific mission and value, as compared to their overall neurologic disease. For the patient to prosper, everyone must strive toward various multidirectional goals that are well orchestrated, noncontradictory, and ultimately effective.

Enhanced decision-making: Healthcare providers may identify areas for improvement and chances for the patient to seize by regularly examining internal (i.e., neurologic function, gut symptoms) and external elements (i.e., new therapy options). This makes it more likely that decision-making will be informed by facts and analysis, rather than intuition.

Improved resource allocation: Under the combined guidance of the healthcare team, patients may distribute their available resources efficiently to get optimal results, by selecting and coordinating their strategies aiming to accomplish an enhanced QoL, despite neurologic deterioration. This guarantees that efforts (i.e., visits) and resources (i.e., medications, interventions) are utilized effectively.

Table 11.4 Strategic management of gut PD

Clear sense of direction (ideally patient-driven)
Enhanced decision-making (based on multifaceted information)
Improved resource allocation
Increased tolerance, adaptability, and long-term adherence
Enhanced physical and cognitive performance
Facilitation of changes in management
Increased use of innovation through up-to-date medical education
Improvement in risk management
Enhanced communication
Build-up of health care team culture

Increased tolerance, adaptability, and long-term adherence: The patient becomes more tolerant and adaptable to challenges that originate from multiple directions (i.e., gut, neuromuscular, cognitive). Through efficient communication, the healthcare provider may recognize opportunities and risks and create strategies to capitalize or avoid them. This improves the patient's physical condition, well-being, and long-term adherence and clinical success.

Enhanced physical and cognitive performance: By setting realistic goals and developing strategies, healthcare providers can ensure that everyone in the team centering upon the patient's QoL is working toward the same aims that are not counterproductive but synergistic. This boosts firm physical and cognitive performance and ensures long-term clinical success.

Facilitation of change in management: Physicians and other members of the healthcare team may identify areas of need and strategize to successfully implement changes, by assessing various internal and external variables. This ensures that the patient can adjust to shifting clinical circumstances and maintain their QoL.

Increased use of innovation: Against an evolving long-term illness, well-informed healthcare providers may promote innovative approaches by examining possibilities in new research and creating plans to take advantage of them. This helps the patient to stay ahead of the disease and succeed in either neutralizing its impact or improving their physical, mental, and emotional performance.

Improved risk management: Regularly assessing various internal and external factors may help the healthcare team identify and mitigate risks. This reduces the consequences of potential threats while ensuring that the patient is prepared for any unexpected events that would in turn compromise their QoL and potentially lead to significant morbidity and mortality.

Enhanced communication: Through their individual perspectives and using effective communication, the various healthcare providers ensure that everyone on the team is working toward the same goals by establishing clear, noncompeting objectives and creating strategies to attain them. This fosters better patient-centered exchange and teamwork.

Buildup of a strong healthcare team culture: Through strategic management in gut PD, a strong healthcare team culture and collaboration can be built, particularly in a specialized, multidisciplinary PD center setting. By defining focused goals, healthcare providers can ensure that everyone is working toward well-aligned aims for the patient.

In summary, a strategic management of patients with gut PD plays an essential role during all stages of the disease. It helps with prevention, recognition, multi-angle initial and longitudinal disease assessment, cooperative setup of direction, decision-making, and resource allocation, thereby boosting patients' QoL, improving physical and mental performance, easing the transition to new ways of doing things, fostering creativity, decreasing anxiety about the unknown, and strengthening team bonding through open lines of communication. Through such management, patients can avoid and respond to shifting disease conditions or complications, by considering all internal and external factors affecting their disease, designing and implementing strategies to meet goals, and monitoring their success over time. This is an ongoing process that requires leadership, clear direction, open lines of communication among providers, and quick thinking to meet shifting and often deteriorating demands of a chronic disease, ultimately aiming at sustaining long-term success.

Strategic management is an iterative process with multiple interdependent steps that, given the evolving nature of the disease, should be periodically reassessed. The major phases are described below. The first phase is the creation of its overall structure, during which a patient and their healthcare team generate a multidisciplinary, thorough, and realistic strategy to seamlessly fulfill all their goals and objectives. After such objectives are established, then strategic options must be identified and assessed for their safety, efficacy, feasibility, and acceptance. Depending on the disease stage and clinical evolution, goal adjustments are made at least once yearly [2].

Management Strategy for Gut PD

There are several key steps in structuring the correct management strategy process, and modifications are frequently needed, given the evolving nature of PD and its underlying gut manifestations (Fig. 11.3).

Define the patient's vision, values, and goals: The patient's vision, values, and goals provide a key framework for strategic planning. The vision is based on the patient's understanding of the nature, severity, and projected natural history of the disease and its specific manifestations, both neurologic (i.e., tremor, rigidity, cognition) and gut-related (i.e., dysphagia, nausea, constipation), and identifies its long-term objectives. The patient's values help mark the significance and impact of the gut disease beyond that of the neurologic manifestations, in turn guiding gut-directed therapy.

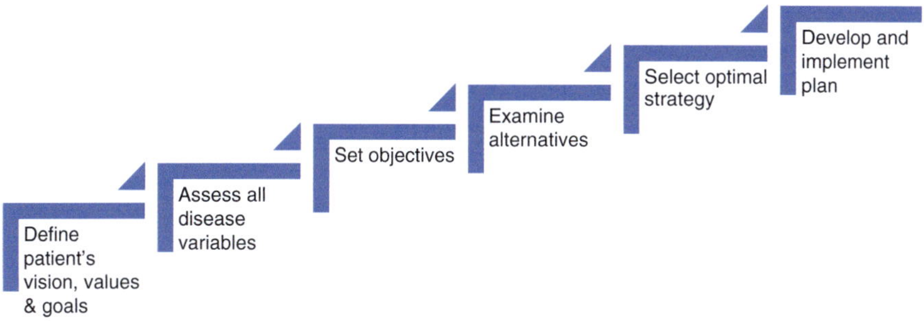

Fig. 11.3 Key steps in structuring the correct management strategy process in Gut PD

Assess all disease variables: An assessment of all disease variables involves looking for any external and internal factors to identify therapeutic opportunities and avoid risks. This requires looking at the speed and severity of disease evolution, patient needs and aspirations, conflicting therapies, disease control, prevailing home and environmental circumstances, home and personal resource availability, physical strength, mobility, and nutritional status.

Set objectives: In setting their objectives based on analysis of all variables, the patient and their care team will also need to consider realistic, organ-focused, and easily quantifiable parameters that will need practical periodic reassessment.

Examine alternatives: The patient and their healthcare team should periodically identify strategic alternatives (i.e., use of endoscopic alternatives for constipation), based on their objectives. This involves generating and evaluating alternative strategies, with their related benefits and risks, that could help the patient achieve their goals. The evaluation and selection of strategic alternatives (i.e., resources, expertise) is always based on their availability, feasibility, suitability, and acceptability and in alignment with the patients' perspectives.

Develop and implement an action plan: This is accomplished by allocating resources, outlining roles, establishing deadlines, and setting up feedback loops to track progress. The plan is then communicated to all who are working toward its achievement. SWOT analysis, a management framework widely used in business, can evaluate the patients' strengths, weaknesses, opportunities, and threats, helping to understand the multitude and sometimes conflicting disease variables and build plans to capitalize on opportunities, rectify weaknesses, and mitigate risks.

SWOT Analysis for Gut PD

Using SWOT, on every patient with gut PD, healthcare providers can highlight existing **strengths**, such as prodromal or early neurological disease, nutritional strength, and preserved physical and mental conditioning; identify **weaknesses**, such as lack of adequate environment, resources, or support, cognitive disabilities, or unwillingness for collaboration with the healthcare team; explore **opportunities** for new medical therapies or interventions, including participation in clinical research trials; and, finally, identify **threats**, such as risk of pulmonary aspiration and malnutrition (Fig. 11.4). Because of the progressive neurodegenerative nature of PD, SWOT analysis needs to be repeated at least yearly so relevant adjustments and preventative steps can be timely implemented [3] (Fig. 11.5).

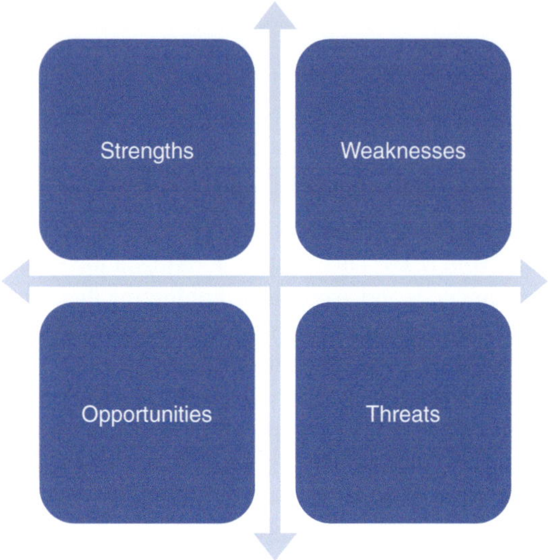

Fig. 11.4 SWOT Analysis for Gut PD

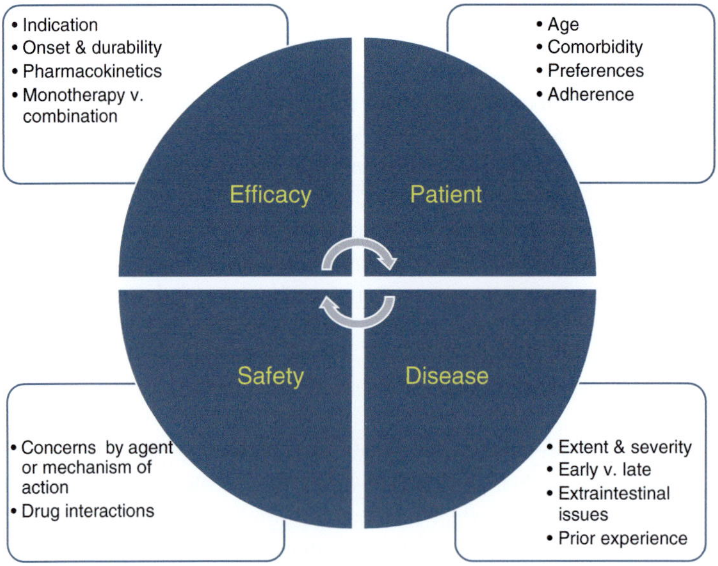

Fig. 11.5 Pieces of therapy choice puzzle in Gut PD

Role of Specialized Centers

Integrated Parkinson's Disease Center (IPDC): An IPDC is a multidisciplinary team with expertise, skill, range, and facilities necessary to achieve optimal outcomes in patients with gut PD efficiently and expeditiously (Fig. 11.6). Within IPDC, patients presenting with symptoms suggestive of gut PD undergo a detailed clinical, functional, and structural evaluation from mouth to anus, prior to implementation of tailored medical, endoscopic, or surgical therapy. Serving as a core, the IPDC clinical practice also supports research and innovation in gut PD, as well as public and physician education. Referrals to the unit may be primary, either from primary care or self-initiated, or secondary from other specialty practices, to reassess patients who have previously failed therapies and to manage complex or complicated cases (Fig. 11.7).

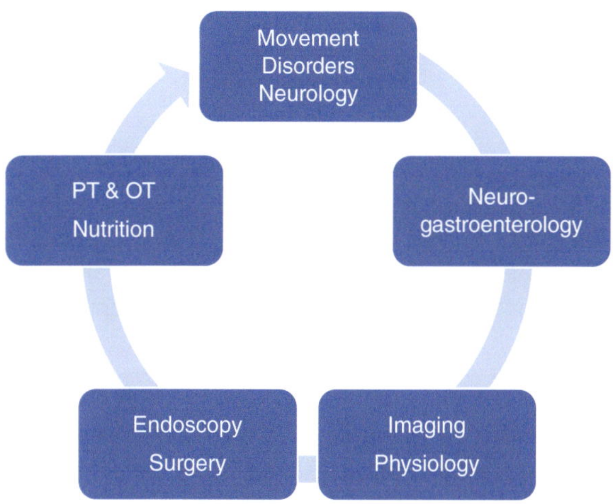

Fig. 11.6 Structure of an Integrated Parkinson's Disease Center (IPDC)

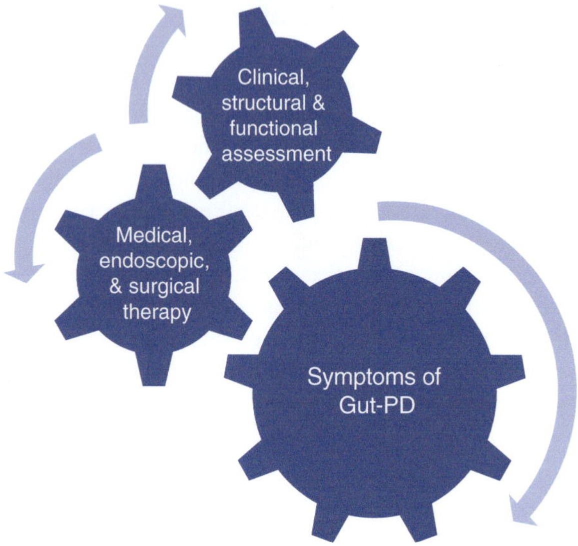

Fig. 11.7 Interactive diagram in assessing and managing Gut PD

Remote Care

Impaired motor and cognitive function can make travel cumbersome for patients with PD, and remote clinic-to-clinic telehealth visits to evaluate the multimodal function in such patients are not only feasible but also help in improving access to care and research participation in such underrepresented populations [4].

References

1. Kübler-Ross E. On death and dying. Scribner; 2014.
2. Voigt I, Fischer S, Proschmann U, Konofalska U, Richter P, Schlieter H, Berger T, Meuth SG, Hartung HP, Akgün K, Ziemssen T. Consensus quality indicators for monitoring multiple sclerosis. Lancet Reg Health Eur. 2024;40:100891. https://doi.org/10.1016/j.lanepe.2024.100891.
3. de-Madaria E, Mira JJ, Carrillo I, Afif W, Ang D, Antelo M, Bollipo S, Castells A, Chahal P, Heinrich H, Law JK, van Leerdam ME, Lens S, Pannala R, Park SH, Rabiee A, Savarino EV, Singh VK, Vargo J, Charabaty A, Drenth JPH. The present and future of gastroenterology and hepatology: an international SWOT analysis (the GASTROSWOT project). Lancet Gastroenterol Hepatol. 2022;7(5):485–94. https://doi.org/10.1016/S2468-1253(21)00442-8. Epub 2022 Mar 2.
4. Virmani T, Pillai L, Smith V, Glover A, Abrams D, Farmer P, Syed S, Spencer HJ, Kemp A, Barron K, Murray T, Morris B, Bowers B, Ward A, Imus T, Larson-Prior LJ, Lotia M, Prior F. Feasibility of regional center telehealth visits utilizing a rural research network in people with Parkinson's disease. J Clin Transl Sci. 2024;8(1):e63. https://doi.org/10.1017/cts.2024.498.

Outcomes Assessment in Gut PD

12

Subjective Assessment

A Likert scale survey question uses a five- or seven-point scale, sometimes called a satisfaction scale, ranging from one extreme attitude to another. Typically, the Likert survey question includes a moderate or neutral option in its scale. Likert scales are easy, practical, and reliable ways to measure patient perceptions about a particular outcome (patient-reported outcome, PRP). Compared to binary questions, which give you only two answer options, Likert-type questions give patients, caretakers, and clinicians more detailed feedback about whether an intervention was just "good enough" or "excellent." Likert questions can help decide whether the intervention left patients feeling "very satisfied," "somewhat dissatisfied," or neutral. This approach uncovers degrees of opinion that could make a real difference in understanding the patient's feedback, pinpointing areas for improvement, or justifying alternative approaches. Patients assess the treatment outcome after therapy using a self-reported global rating of change in overall symptoms on a seven-point Likert scale (how would you describe your bowel condition *now* compared with how it was *before* you started treatment? Select from: (1) very much worse; (2) much worse; (3) worse; (4) no change; (5) slightly better; (6) much better; and (7) very much better). Such seven-point Likert scales are sensitive to small and potentially significant responses to treatment, with a rating of 6 or 7 considered a clinically meaningful improvement. Patients had to achieve these scores to be considered responder to therapy. Likert scale ratings are structured to provide quantifiable answer options that make analyzing data easier. Respondents also have a range of answers that are more specific to how they feel about an intervention. Likert scales work better when questions are focused, avoiding broad questions that respondents may find too hard to think about. This could lead them to get frustrated and start answering too quickly, spoiling data quality [1].

G. Triadafilopoulos, *The Gut in Parkinson's Disease*,
https://doi.org/10.1007/978-3-031-77171-2_12

Objective Assessment

Depending on the organ affected in gut PD, objective measurements of clinical parameters can be made before, during, and after any intervention, be it pharmacologic, endoscopic, or surgical. Such clinical parameters however will need to reflect to the highest possible degree the patients' symptoms. For example, measurement of gastric emptying rate (in hours) is known to correlate poorly with symptoms in patients with gastroparesis, and its use as a surrogate marker of clinical disease activity may be unreliable. Similarly, a poor correlation exists between stool burden assessed by imaging and symptomatic magnitude of constipation in gut PD. Prospective, intermediate-term studies (up to 3 years) have nevertheless shown stability of clinical motility test parameters, such as strength of esophageal peristalsis, upper or lower esophageal sphincter pressure, internal and external anal sphincter pressures, and anorectal coordination in patients with gut PD. Therefore, such clinical measurements could longitudinally serve as markers for disease progression, either naturally or as reflection of an intervention.

Combination Assessment

Although not formally assessed in gut PD, it is possible that a combination assessment of both subjective measures, using Likert scales or even more sophisticated and validated disease questionnaires with objective tools, such as motility tests, could provide a better validation of disease activity and symptom severity, but this has not been tried in gut PD. More research is needed in this regard.

Reference

1. Xu J, Wang J, Keith S, Zhang M, Yang C, Yuan Q, Qiu Y, Hu X, Wu X. Management of Parkinson's disease patients after DBS by remote programming: preliminary application of single center during quarantine of 2019-nCoV. J Neurol. 2021;268(4):1295–303.

Research in Gut PD

13

Research in PD

Prevention of PD is an important focus of research to address genetic changes and disparities across sex, race, ethnicity, socioeconomic status, lifestyle modification, and reduction of exposure to environmental toxins. Telemedicine can improve access to care, while wearable digital devices and virtual reality may improve screening, monitoring, and treatment [1]. Biomarkers for abnormal α-synuclein can distinguish persons with PD, dementia with Lewy bodies, or REM sleep behavior disorder from healthy controls and from persons with other neurologic diseases, with high sensitivity and specificity. Testing of α-synuclein may allow early detection of neuronal α-synuclein disease, which could in turn lead to early intervention [2].

Therapeutic Modulation of the Gut Microbiome

Presently, it is unclear whether the microbiome alterations are causal or just correlative and secondary or shared nonspecific or even compensatory responses to disease. Therefore, therapeutic manipulation of the microbiome could be potentially detrimental. Prebiotics, mostly nondigestible carbohydrates, are substrates that are utilized by host microorganisms to confer health benefits. In contrast, probiotics are live microorganisms that confer health benefits when taken by mouth. Fecal microbiota transplantation involves transferring liquid fecal filtrate containing a mixture of microbes from a healthy individual to a patient through nasogastric or nasoduodenal tube or during colonoscopy. Postbiotics are preparations of killed microorganisms that confer a health benefit. Their effects are due to inhibition of pathogen colonization, modification of the microbiome composition or function, reduction of inflammation, modulation of local immunity, and improvement of host gut epithelial barrier function.

G. Triadafilopoulos, *The Gut in Parkinson's Disease*,
https://doi.org/10.1007/978-3-031-77171-2_13

The use of antimicrobials reduces the numbers of many gut microbes without distinguishing between harmful and beneficial ones. Although transiently effective, repeated courses of treatment for small intestinal bacterial overgrowth carry a risk of antibiotic resistance. Targeted elimination of specific bacterial strains using bacteriophages holds promise for the future.

Small molecules, biologics, or metabolites that act on gut-related targets could include immunomodulatory therapies, short-chain fatty acids, and gut barrier-restoring treatments such as tight junction modulators or mucus stabilizers. Many of these treatment modalities can be personalized, based on diet and medication intake and microbiome, metabolome, and/or immune profiles, to optimize treatment success.

Enhancing Gut Motility and Modifying Sensation

At best, the clinical effectiveness and safety of various prokinetic drugs in gut PD today is unclear since there are limited studies focusing on this population. There is also a large unmet need for development and clinical application of both regionally selective and globally acting prokinetic drugs that work primarily intraluminally and are safe and without systemic side effects in the general population as well as in those with gut PD. Further research will also be needed to explore optimal dosing regimens, long-term safety profiles, and potential combination therapies of prokinetic drugs to enhance patient care and outcomes [3].

Since the gastrointestinal tract has its own independent nervous system, known as the enteric nervous system (ENS), advances are made on how neurons within the ENS coordinate sensory and motor functions. Knowledge of how distension or chemical stimulation of the intestine evokes sensory responses in the ENS and central nervous system has progressed, including critical elements that underlie the mechano-transduction of distension-evoked colonic peristalsis. It seems that mucosal serotonin is not required for peristalsis or colonic migrating motor complexes and that chemosensory stimuli applied to the lumen can release substances from enteroendocrine cells, which could subsequently modulate ENS activity. Further work will be needed to enhance our understanding of how intrinsic sensory neurons in the ENS detect and respond to sensory stimuli and how these mechanisms differ from extrinsic sensory nerve endings in the gut that underlie the gut–brain axis [4].

Medical Devices

There is significant potential for the development and clinical implementation of medical devices to assist patients with gut PD to improve their quality of life. Devices to treat constipation have been developed or are under development but have not been studied in PD cohorts [5]. An intestinal sampling device (CapScan) is already clinically available for research purposes, and it promises not only to sample the intestinal microbiome and other potential contributors to disease pathophysiology but also deliver intraluminal therapeutic molecules in gut PD [6]. Further exploration of the potential of medical devices in gut PD will require the close collaboration between neurogastroenterologists and movement disorder specialists.

References

1. Cubo E, Delgado-López PD. Telemedicine in the Management of Parkinson's disease: achievements, challenges, and future perspectives. Brain Sci. 2022;12(12):1735. https://doi.org/10.3390/brainsci12121735.
2. Siderowf A, Concha-Marambio L, Lafontant DE, Farris CM, Ma Y, Urenia PA, Nguyen H, Alcalay RN, Chahine LM, Foroud T, Galasko D, Kieburtz K, Merchant K, Mollenhauer B, Poston KL, Seibyl J, Simuni T, Tanner CM, Weintraub D, Videnovic A, Choi SH, Kurth R, Caspell-Garcia C, Coffey CS, Frasier M, Oliveira LMA, Hutten SJ, Sherer T, Marek K, Soto C. Parkinson's progression markers initiative. Assessment of heterogeneity among participants in the Parkinson's progression markers initiative cohort using α-synuclein seed amplification: a cross-sectional study. Lancet Neurol. 2023;22(5):407–17. https://doi.org/10.1016/S1474-4422(23)00109-6.
3. Bor S, Kalkan İH, Savarino E, Rao S, Tack J, Pasricha J, Cangemi D, Schol J, Karunaratne T, Ghisa M, Ahuja NK, Lacy B. Prokinetics-safety and efficacy: the European Society of Neurogastroenterology and Motility/the American Neurogastroenterology and motility society expert review. Neurogastroenterol Motil. 2024;36(5):e14774.
4. Spencer NJ, Hu H. Enteric nervous system: sensory transduction, neural circuits and gastrointestinal motility. Nat Rev Gastroenterol Hepatol. 2020;17(6):338–51.
5. Uwawah TD, Nduma BN, Nkeonye S, Kaur D, Ekhator C. A novel vibrating capsule treatment for constipation: a review of the literature. Cureus. 2024;16(1):e52943.
6. Shalon D, Culver RN, Grembi JA, Folz J, Treit PV, Shi H, Rosenberger FA, Dethlefsen L, Meng X, Yaffe E, Aranda-Díaz A, Geyer PE, Mueller-Reif JB, Spencer S, Patterson AD, Triadafilopoulos G, Holmes SP, Mann M, Fiehn O, Relman DA, Huang KC. Profiling the human intestinal environment under physiological conditions. Nature. 2023;617(7961):581–91.

Conclusions and Key Messages 14

PD is a progressive disorder of later life that is defined clinically by motor (bradykinesia, rigidity, tremor, and imbalance) and non-motor features (hyposmia, autonomic dysfunction altered cognition).

PD is pathologically characterized by neuronal degeneration and intraneuronal misfolded α-synuclein (Lewy bodies) in specific central and peripheral nervous system regions, including dopaminergic brainstem neurons.

Mood disorders, sleep, sensation, cognition, and symptoms of autonomic dysfunction often precede motor signs by years (prodromal PD) and worsen with the duration of the disease.

More effective and PD-specific treatment modalities are needed. No present therapy can slow progression. Dopaminergic therapies improve motor function, but progressive loss of efficacy and side effects are common.

Non-motor symptoms, such as gut PD, cause substantial morbidity and respond to off-label medications and other therapies. Comprehensive multidisciplinary care is important.

Deep brain stimulation surgery (DBS) is effective for motor fluctuations, but evidence of efficacy against gut PD is lacking.

Understanding of the key role of the gastrointestinal system in PD has evolved substantially during the past two decades. Although progress has been made in our management of gastrointestinal dysfunction in these patients, much more needs to be accomplished.

The potential roles of *H. pylori* infection and SIBO in the development of gastrointestinal dysfunction in PD might allow new treatments.

Research into other ways to administer levodopa bypassing the gastrointestinal tract is also being explored.

The role of probiotics aiming at altering the gut microbiome, preventing SIBO, or altering the progression of the neurological manifestations is under present evaluation.

High-resolution esophageal manometry helps in the management of dysphagia.

G. Triadafilopoulos, *The Gut in Parkinson's Disease*,
https://doi.org/10.1007/978-3-031-77171-2_14

The gastroparesis seen in PD may contribute not only to symptoms (bloating, weight loss) but also to poor drug efficacy.

Glucose or lactulose breath testing detects small intestinal bacterial overgrowth and guides therapy.

Most PD patients have multifactorial constipation and will need therapy for both defecatory dyssynergia and colonic inertia.

Fiber is not well tolerated by patients with colonic inertia and is unlikely to be useful in most patients with PD.

Since rectal sensation is frequently preserved in gut PD, bulking with fiber—aimed at improving rectal sensation—may not be needed and may not be well tolerated.

Pelvic floor physical therapy should be tried in PD patients with defecatory dyssynergia, but its efficacy is unknown.

Osmotic laxatives, linaclotide, lubiprostone, and prucalopride could be useful in the treatment of colonic inertia in PD, since they shorten colonic transit time and mitigate fecal impaction.

Since many PD patients with constipation have low anal pressures, the occurrence of fecal incontinence may be limiting.

PD patients with absent RAIR may need programmed defecation.

Injection Botox therapy may be useful in some PD patients with dysphagia, bloating, or constipation.

Index

G. Triadafilopoulos, *The Gut in Parkinson's Disease*,
https://doi.org/10.1007/978-3-031-77171-2

MIX
Papier aus verantwortungsvollen Quellen
Paper from responsible sources
FSC® C105338
FSC
www.fsc.org

If you have any concerns about our products,
you can contact us on
ProductSafety@springernature.com

In case Publisher is established outside the EU,
the EU authorized representative is:
Springer Nature Customer Service Center GmbH
Europaplatz 3, 69115 Heidelberg, Germany

Printed by Libri Plureos GmbH
in Hamburg, Germany